The Illustrated Encyclopedia of
HELICOPTERS

ERA
N351EH

The Illustrated Encyclopedia of
HELICOPTERS

Michael J Taylor

Exeter Books
NEW YORK

Photographic acknowledgments
Aérospatiale 96 bottom, 98, 99, 161 bottom, 166 bottom, 167 bottom, 176; Agusta 152, 161 top; Avia Export 108, 109 top; Aviation Photographs International/Jeremy Flack 145 bottom; Bell-Boeing Vertol 146 bottom; Bell Helicopter Textron 51, 52, 57, 59, 115, 118, 119, 124, 125, 126, 128 top, 130, 132, 133, 134, 136 bottom, 160, 162, 163, 164, 169, 170-171, 172, 173 top, 174; BIIL 26 bottom; Peter Bish 144 top; Boeing Vertol 54, 55 bottom, 79 bottom, 80, 81 top, 82 bottom, 83 top, 113, 116, 117 top, 121 bottom, 122 bottom, 123, 129, 173 bottom; Breda Nardi 72 bottom; British Airways 61 bottom; Austin J. Brown 24; ECP Armées 97; Flight International 21, 94 bottom; Hiller Aircraft Corporation 66 top; Hiller Aviation 167 top; Hiller Helicopters 64 top, 114; Bill Hobson 19; Hughes Helicopters 7, 9, 71 bottom, 72 top, 127, 137, 138-139, 140, 141, 142 top, 146 top, 165, 166 top; Leslie Hunt 47; Phillip Jarrett 36; Kaman Aerospace 67, 68 bottom, 69; Kawasaki 175 bottom; Lockheed 134-135 top; MBB 142 bottom, 143 top, 168; McDonnell Douglas 73 top; T. Melnib 109 bottom; NASA 153; New York Airways 82 top; Novosti 104; S.P. Peltz 105 bottom; Piasecki Aircraft 84 bottom; Pilot Press 44; Pratt & Witney 46; PZL-Swidnik 106 bottom; RAF Museum, Hendon 30, 60 bottom, 94 top; Royal Danish Ministry of Foreign Affairs 16 bottom; Sabena 58 top, 75 bottom; Science Museum, London 6; Brian M. Service 185; Sikorsky Aircraft 20, 38, 48 bottom, 50 bottom, 73 bottom, 74 top, 76 top and centre, 79 top, 121 top, 147, 148, 150 bottom, 154, 155, 156, 157, 158-159; J.W.R. Taylor 31; United Helicopters 53; US Air Force 41, 42, 45, 49, 70 top right, 83 bottom, 106 top, 145 top; US Army 85 top; US National Archives, Washington D.C. 71 top, 110, 111, 117 bottom, 120, 122 top, 131; US Navy 70 top left, 105 top, 182; Wg. Cdr. K.H. Wallis 29; Mick West 184-185; Westland Helicopters 22 bottom, 58 bottom, 61 top, 74 bottom, 75 top, 77 bottom, 78, 88 top, 89, 91, 92 top, 93, 112 bottom, 143 bottom, 149, 150 top, 177, 178-179, 180, 181.

All other photographs from the collection of Michael J. Taylor.

Front cover: Sikorsky UH-60A Black Hawk, carrying 16 Hellfire missiles on its ESSS (External Stores Support System).

Back cover: Sikorsky S-76 Mark III commercial and corporate transport helicopter.

Titlespread: A 5/6 seat AS 350D Astar belonging to ERA Helicopters.

First published in the USA 1984
by Exeter Books
Distributed by Bookthrift
Exeter is a trademark of Simon & Schuster, Inc.
Bookthrift is a registered trademark of Simon & Schuster, Inc.
New York, New York

Prepared by
Deans International Publishing
52-54 Southwark Street, London SE1 1UA
A division of The Hamlyn Publishing Group Limited
London · New York · Sydney · Toronto

ISBN 0 671 07149 1

Printed in Italy

Contents

Windmills and Whirligigs

The helicopter in its most basic form predates gliders, powered aeroplanes, balloons and airships. Only the earliest forms of wings, such as those used by the so-called 'tower jumpers' who leapt from buildings and walls to death or injury, and early forms of the rocket precede the helicopter in the annals of aviation history. But the helicopter as a man-carrying aircraft is an invention of the 20th century: its practical development beyond theory proved far more difficult than for other forms of flying machine.

The word *helicopter* is derived from the Greek '*helix*' and '*pteron*', meaning 'spiral' and 'wing', but the actual historical events that shaped the conception of the helicopter occurred in 7th century Persia. To Persia is attributed the first mention of windmills. Those known to have been constructed in the 10th century, used horizontally rotating sails. This idea quickly spread, first to China where Persians, captured by the warlord Gengis Khan, put their knowledge into practice. Europeans invented the tower form of vertical windmill in the 12th century – the amalgamation of simple wind-power technology brought back from the medieval holy crusades in the East and of existing watermills. France appears to have been first to erect a European windmill (AD 1180), followed some years later by Britain.

Although there is little basis for certainty, it seems entirely probable that a hand-held whirling sail toy, a primitive version of that still much-enjoyed by children, was circulating later the same century. From this simple plaything a more important toy developed. By adopting a method of mechanically rotating the central shaft using a pull-string or bow drill (itself an invention of man's Upper Paleolithic period and thus long established and understood), the toy's sails could be made to spin and even fly away. The first known illustration of such a toy is in a Flemish manuscript *circa* 1325, and a French painting of 1460 depicts a child holding a fairly sophisticated pull-string helicopter toy.

Later the same century the genius Italian inventor, philosopher and artist Leonardo da Vinci designed a form of helicopter with a corkscrew spiral rotor made from starched flaxen linen. This, he considered, would 'screw' its way

Below: Leonardo da Vinci's design for a helicopter, intended to 'screw' its way upward using a spiral rotor made from starched flaxen linen.

Below right: Hughes NOTAR helicopter, flown in December 1981. A modified OH-6, the tail rotor has given way to slots through which pressurized air is vented.

upward under power. Whether or not da Vinci actually constructed a model helicopter of this design has not been established. It is known, however, that among his ideas for providing power for aircraft was one in which the pilot continuously pulled and then rewound a rope round a central shaft mechanism in a typical string-pull manner. Da Vinci had clearly taken notice of the toy's power source and, interestingly, it was his use of Greek that gave the world the word 'helicopter'.

As it is unlikely that da Vinci built a model or full-size version of his helicopter, he did not come face to face with an inherent design fault in his machine. Helicopters have to counter what is known as the torque effect, which simply means that when a rotor is turned by the power from a structure-mounted engine, the engine and carrying structure spin in the opposite direction to the rotor. Once this effect was understood, methods were employed to counter torque and thus allow only the rotor to spin. The most common early method was to use two main rotors that turned in opposite directions (contra-rotated) and so countered each other's torque effect; most later-developed helicopters used a tail rotor to produce thrust in the direction opposite to the torque effect. The very latest method, under test in the U.S.A. in the 1980s calls for the removal of the tail rotor on single-rotor helicopters. It is replaced by a system of pressurized air which is vented from a tailboom slot which, when pushed downward by the downwash from the rotor, exerts an anti-torque force as it passes over part of the tailboom. This countering method may become particularly relevant for future high-manoeuvring helicopters or those requiring low-noise characteristics.

What experimentation in the field of helicopters was undertaken between da Vinci's death in 1519 and the 18th century is not well documented but the second half of the 18th century witnessed some interesting developments. It is now thought possible that a Russian named Mikhail Vasilyevich Lomonosov may have flown a model helicopter in 1754. If so, the Lomonosov ousts the Frenchmen Launoy and Bienvenu as the first to fly a self-propelled model helicopter and, indeed, the first to fly a powered aircraft of any description if one (rightly) excludes the rocket. Lomonosov (November 1711-April 1765) is famed in Russian history as a man of science and literature. He was responsible for founding a university in Moscow and ran a glass works from which glass mosaics originated. He also expounded his theories on electricity and, as mentioned above, may have flown a simple vertical-lift model helicopter with contra-rotating rotors.

However, the Frenchmen Launoy and Bienvenu are more generally recognized as having flown the first model helicopter, although this recognition may be due entirely to the fact that exact dates, places and descriptions surrounding this milestone event are well documented. A naturalist and a mechanic respectively, the Frenchmen demonstrated the helicopter on 28 April 1784 at the Académie des Sciences in Paris, France. This model also had contra-rotating rotors, each rotor covered with fabric and attached to the ends of a stick. Power was provided by a bow drill, the string of which had been wound round the stick in such a way as it would unwind upon release.

In Britain in 1796, Sir George Cayley, known to aviation historians as the 'Father of Aerial Navigation' because of his remarkable work in the field of heavier-than-air flight, constructed and flew a model helicopter which was virtually identical to that of the Frenchmen, except that he adopted four-blade rotors made from bird feathers. It is unlikely that Cayley had any knowledge of the French model. Cayley, a Baronet living at Brompton Hall near Scarborough, England, died in December 1857, well before any man-carrying helicopter flew. However, designs for man-lifting helicopters were produced in the 18th and 19th centuries and Cayley himself designed a machine with a boat-like wheeled fuselage incorporating a bird's head bow, two rear-mounted four-blade propellers for horizontal propulsion and four rotors. The rotors are of particular interest as they each comprised eight blades that pivoted into flat wings for horizontal flight when vertical lift was not required. An aircraft of this design would today be termed a 'convertiplane'.

An earlier design for a full-size helicopter, and one which was actually built and tested, was the brainchild of Jean-Pierre Blanchard. This Frenchman later became a renowned balloonist and his feats in the field of lighter-than-air flying included the first use of a propeller fitted to a balloon basket in an attempt to propel the craft (1784), the first air crossing of the English Channel (by balloon in 1785), and a daring parachute descent (1793). Blanchard constructed a man-powered helicopter in 1782, but unfortunately it failed to lift from the ground. However his aspirations to become airborne were fulfilled by the hydrogen balloon.

Aviation history has to jump to the year 1828 before any real progress could be documented, almost certainly because the hot-air and hydrogen balloon already gave would-be aviators a proven, simple and relatively safe way of becoming airborne without involving the crew in any strenuous actions to generate lift through mechanical systems. However, in 1828 Briton David Mayer built and tested an unsuccessful helicopter which again required the muscle power of its would-be pilot. At the same time in Italy contra-rotation was the subject of consideration by Vittorio Sarti.

An ingenious method of getting round the problems associated with mechanically turning a rotor was found by Englishman W. H. Phillips, who propelled the rotor of a model helicopter by gas forced out of the rotor tips. His model of 1842 is said to have flown a considerable distance and is remembered as the precursor of the pressure-jet helicopter, examples of which appeared and were flown as full-size helicopters from the mid-20th century. A reconstruction of Phillips's helicopter was one exhibit at the first-ever aeronautical exhibition, staged in 1868 by the Aeronautical Society of Great Britain at Crystal Palace.

In 1843 an Englishman by the name of Bourne took the earlier idea of using feathers as rotor blades a stage further and flew a number of tiny helicopters powered by watch springs. Two years later a designer by the name of Cossus proposed a three-rotor helicopter using a steam engine as the power source. This craft was never built and, as the imaginative William Henson found out about the same time when attempting to achieve a sustained flight with his 20-foot (6.1-metre)-span fixed-wing model aircraft, had it been constructed it would not have achieved flight owing to the heavy weight of steam engines of the day. The most important aspect of Cossus's theoretical work was that he was the first to mention in specific terms the idea of pivoting rotors in the direction in which one intended to progress.

Among those who produced helicopter designs in or about the middle of the 19th century was the Briton Henry Bright, who clearly mistrusted all available forms of engine and took the rather backward step of designing a man-powered machine. In France, which took the greatest interest in helicopters of any European nation and later was rewarded with the first man-carrying helicopters to lift from the ground, Vicomte de Ponton d'Amécourt attempted to fly a model helicopter in 1863. This well-conceived model was a failure in terms of flight because of the steam engine selected but otherwise was a model of some merit. Supported on three legs, the helicopter used a steam engine to power two contra-rotating, paddle-like, two-blade rotors, the shaft for the upper rotor passing through the tubular shaft of the lower. As if to prove that it *was* the weight of the steam engine that prevented the model from flying, he thereafter shifted to clockwork power for his models, a change that worked well for tiny models but which could not be translated to a full-size helicopter.

A man who contributed little to the advancement of helicopters but is acknowledged as one of the most outstanding figures of early aeronautics in general was the Frenchman Alphonse Pénaud. A tragic figure from a naval family, he turned to aeronautics at the age of 20 and took his own life a decade later in 1880. Among his first experiments were those relating to twisted rubber-powered model helicopters, which flew well. However, he is better remembered for his rubber-powered model aeroplane of 1871 and a similarly powered flapping ornithopter of 1874 that often covered a distance of 100 feet (30 metres). The recognized peak of his aeronautical achievements was the design of a full-size powered aeroplane, patented by himself and his mechanic Paul Gauchot on 18 February 1876. This aeroplane had what was basically a reverse delta wing with dihedral, incidence and camber; an enclosed fuselage for the pilot which doubled as a flying-boat hull for water operation; a tail unit with fin, rudder and elevator surfaces; tractor-mounted propellers; and a fully retractable undercarriage. Had Pénaud lived longer it is doubtful that he would have returned to the helicopter and one can only speculate what he might have achieved had he done so.

Although Pénaud's involvement with helicopters was fairly unspectacular, other engineers, inventors and designers researching along this line were also only achieving success with models. However, a design for a large steam-powered helicopter that emerged from the pen of Achenbach in 1874 incorporated, for the first time, an anti-torque secondary rotor. Two years later the American J. Ward produced the design for a machine which encompassed the technologies of the helicopter and hovercraft.

The year 1877 saw the clever use of steam for model helicopters, with Italian Enrico Forlanini and Frenchman Emmanuel Dieuaide dispensing with heavy water-heating apparatus on their craft. Forlanini removed the boiler from the engine, heated it to pressure, and only then attached it to his helicopter. Dieuaide went one stage further by leaving the boiler with the heater and feeding steam to the model by pipe. Dieuaide had little success with his helicopter but Forlanini's, also featuring contra-rotating rotors, could stay airborne for a brief but impressive period. In 1878 another Frenchman, Castel,

flew a model helicopter with contra-rotating rotors, using compressed air fed into the machine in a similar way to the steam on Dieuaide's helicopter. However, while it can be claimed that this was the first large model helicopter with a pressure-jet propulsion system, it proved unstable and was damaged beyond repair before any significant flight was made. It is interesting to note that much later pressure-jet propulsion systems using compressed air were widely experimented with, particularly during the 1950s, and that the helicopter with the largest rotor ever, the American Hughes XH-17 with a rotor diameter of 130 feet (37.62 metres), used such a system.

One further name should be mentioned in connection with early helicopter experiments. In 1884 Charles Parsons of Great Britain designed what is today recognized as the first practical steam-turbine engine. An inventor of many interests, he constructed a helicopter in 1894 with a tiny steam engine, which he subsequently modified into an aeroplane. Although in aviation history this experiment is of little importance, it is interesting to note that nine years later Parsons took onto his payroll a certain gentleman named Horace Leonard Short as an experimental engineer. Horace Short subsequently became one founder brother of the Short Brothers aircraft company, the first aircraft manufacturer in the world to put an aeroplane into series production.

France takes the initiative

The dramatic success of the American Wright Brothers with *Flyer*, their powered aeroplane, at the end of 1903, was barely acknowledged internationally until more substantial flights were achieved in the following year. However, in November 1904 Wilbur Wright flew his *Flyer II*

Hughes XH-17, with the largest rotor of any helicopter ever.

over a distance of 2.75 miles (4.43 km) at Dayton, Ohio, and the world at last paid full attention to the aeroplane. In January of the following year the American government began discussions on the possibility of purchasing a Wright biplane as a military aircraft. From 1905 notable events with aeroplanes came thick and fast. In October 1905 Wilbur Wright made a sustained flight of about 38 minutes duration. In November 1906 the first officially recognized sustained flight in Europe was made by Alberto Santos-Dumont in France, which had become the most air-minded nation in Europe, if not the world. France had also been regarded as the greatest exponent of the helicopter, although there was little to show for the long period of experimentation.

The birth of the aeroplane proper should have made experiments in other fields of powered heavier-than-air flying more credible but virtually the opposite became the case. Many engineers still viewed the helicopter as a machine that offended all the finer laws of aerodynamics. The last to condemn helicopter research should have been Wilbur Wright who, with his brother Orville, had himself worked on helicopters before changing to aeroplanes and had, early in his career, been made all too aware of sceptics in aviation. Yet, in early 1906, Wilbur wrote a letter to a colleague in which he said, 'Like all novices we began with the helicopter (in childhood), but soon saw that it had no future and dropped it. The helicopter does with great labor only what the balloon does without labor. . . . The helicopter is much easier to design than the aeroplane but it is worthless when done.'

It was an unfortunate fact that before *any* manned heavier-than-air machine flew properly, most persons engaged in aeroplane experimentation were looked upon with sceptical amusement. But once a form of powered flying machine had been demonstrated successfully, it was thought that anyone still researching in other fields *had* to be misguided. As Wilbur had, by 1906, already flown about 24 miles (39 km) and no

One of the eight-blade rotors of the Breguet-Richet Gyroplane No. 1. *The 50-hp Antoinette engine is seen to the right of the rotor.*

Breguet-Richet Gyroplane No 2, *a hybrid aeroplane and helicopter.*

piloted helicopter had yet proved capable of lifting off the ground, such thoughts appeared reasonable. Of course Wilbur was incorrect in his conclusion that the helicopter was easier to design. Indeed, the opposite was to be the case.

The Wright brothers had achieved a successful aeroplane design only after years of hard research and experimentation. At the same time others, attempting to flap their way skyward in ornithopters or fly in illbegotten fantastic machines of all shapes and sizes built without sound theoretical basis, had served only to heap ridicule on the whole movement. Unfortunately, the many ill-conceived, mechanically unsound and aerodynamically hopeless full-size helicopters built around the turn of the century had a similar effect. Interspersed with these, however, were a number of better designs which, though incapable of flight, were the result of knowledgeable research. One such worthy machine was the *Bremen 1*. This remarkable machine took Carl Zenker many years to design, develop and build, and adopted no fewer than four pairs of four-blade rotors for lift and two propellers for horizontal propulsion, all carried on a framework of bamboo. Tested in early 1900, it failed to lift but was not unlike, in conception, the first manned helicopters to achieve a measure of flight.

The first manned helicopters to fly had no practical use in themselves but were responsible for proving that a full-size rotary-wing machine could lift its weight plus that of a pilot off the ground. As seems appropriate, both of the pioneering helicopters in this respect were of French origin. The first to lift from the ground was the massive structure designed by Louis Breguet and Professor Richet. Known as the Breguet-Richet *Gyroplane No. 1*, it was basically a cruciform structure with a wheel at the end of each boom, a 50-hp Antoinette eight-cylinder inline engine at the centre under which the pilot sat, and a 26-foot 3-inch (8-metre) rotor had eight blades above each outer wheel, making four rotors in all. The complete structure with pilot weighed a massive 577 kg (1,272 lb). On 29 September 1907 *Gyroplane No. 1* lifted itself at Douai but proved unstable and had to be steadied in flight by four persons using long poles.

Having achieved success with *No. 1*, Breguet and Richet designed a more sophisticated helicopter. This became known as the Breguet-Richet *Gyroplane No. 2* but was a hybrid aeroplane and helicopter. *No. 2* used a covered aeroplane-type fuselage, fitted with a tailplane and rudder. Between the biplane wings, spanning 32 feet 10 inches (10 metres), were carried two *Gyroplane No. 1*-type rotors, giving the aircraft an overall width of 45 feet

Paul Cornu's twin-rotor helicopter, recognized as the first manned helicopter to make a free flight.

11 inches (14 metres). Each rotor was inclined forward at 40 degrees for rising and the machine incorporated forward elevators. However, the third aircraft by Breguet and Richet was an aeroplane and Breguet thereafter specialized in aeroplanes. Indeed, in March 1911 Breguet flew one of his aeroplanes with 11 passengers on board over a distance of 3 miles (5 km), the first time such a number had been lifted, and his warplanes achieved good results during the 1914–18 war. Breguet aeroplanes continued to be built after the war and a further Breguet helicopter worthy of note did not appear until the 1930s. This is mentioned in a later chapter.

The second helicopter ever to lift from the ground carrying a man was designed and built by Paul Cornu. This Frenchman had earlier built and tested a model helicopter powered by a 2-hp Buchet engine and it was upon the results of these tests that he developed his full-size machine. The full-size twin-rotor helicopter was completed in August 1907. The main structure comprised a 'Vee' of large-diameter steel tubing, attached to which was a preponderance of small-diameter tubing and bracing wires. On the flat centre portion of the main 'Vee' was fitted a 24-hp Antoinette engine and the seat for the pilot, the whole resting on a four-wheel undercarriage. At the extreme ends of the main 'Vee' were carried two horizontally mounted 5-foot 9-inch (1.77-metre) wheels, upon which the two-blade rotors were fixed. The wheels were turned by a single wide belt-drive from an engine shaft, a system which was less successful on this helicopter than on the earlier model. Further out from the wheels but under the rotor blades were fore-and-aft elevators, which had several functions in controlling the helicopter. By using two levers, Cornu could control these surfaces in all directions, allowing the downwash from the

The first French Bertin contra-rotating rotor helicopter.

rotors to be directed as required, thus providing not only control of direction but speed.

As a wise precaution, Cornu first flight-tested his helicopter with a sack of soot representing the weight of the pilot. On 27 September 1907 it rose from the ground, two days before *Gyroplane No. 1*. However, it was not until 13 November that year that Paul Cornu piloted the helicopter as it lifted on its 20-second duration first manned flight at Lisieux. Although the altitude attained was a mere 1 foot (0.3 metres), the helicopter was not steadied in flight and has been adjudged, therefore, to be the first helicopter in aviation history to make a free flight. The irony is that Cornu's helicopter was tethered for this flight as a precaution against an uncontrollable surge upwards. However, unlike some aeroplanes whose tethered flights have not been accepted as true flights from an historical standpoint, the tether line to Cornu's helicopter played no part in the course of the flight and so is generally overlooked. This helicopter, each rotor of which had a diameter of 19 feet 8 inches (6 metres), weighed 573 lb (260 kg) for this historic flight.

Experimentation gathers pace

Following the achievements in France in 1907, a steady stream of helicopters appeared over the ensuing years in several countries. Some of these had merit. Historically, few are worth mentioning, although some reference should be given to those that furthered the development of the practical helicopter. One or two of the more bizarre are also included.

In Britain, Howard Wright built two machines for an Italian named Cappone. Wright attempted to combine the features of a glider and a twin-rotor helicopter and so provide machines which could stay aloft for reasonable periods on low power. Known as helicoplanes, large sums of money were spent on their development but they were complete failures. In France, Bertin produced first a conventional contra-rotating helicopter, followed in 1908 by a machine which combined an aeroplane's configuration with a small rotor, a 55-hp engine providing power for both the tractor propeller and rotor. Also in 1908, the German Degn designed a machine in which a Daimler engine powered a rotor and flapping wings, apparently attracting some attention as a possible aircraft for use on board warships. In the U.S.A., the birthplace of the powered aeroplane proper, many helicopters appeared *circa* 1909–10. The English machine, built in San Francisco and initially tested in its shed in July 1909, used a 60-hp engine to drive two rotors, each of which had two disc-type blades. This machine, apparently, broke loose during tethered trials and was severely damaged. Disc rotor blades were used also by the Irvine helicopter, while the rather primitive contra-rotating Luyties Otto helicopter was reported to have risen from the ground on several occasions. An award for just trying should have been given to the Rickman machine, whose large umbrella-type rotor was driven by two men pedalling hard on a tandem tricycle. Yet another contra-rotating machine was the Williams helicopter.

In Russia several helicopters were built *circa* 1910. The one developed by Sikorsky is mentioned in a later chapter. However, the Russian government was interested in furthering the progress of helicopter design and it is said that five helicopters came from Baranovski alone. A contemporary report suggests that one of these, which also had monoplane wings, left the ground very rapidly in vertical flight and was officially selected for further research.

Flapping hinges and cyclic pitch control

It has not always been possible for historians to give credit for innovations in helicopter design, as early development was, to some degree, obscured by the more rapid progress achieved by fixed-wing aircraft. However, certain names have been recorded over the years, such as that of Achenbach (previously mentioned) who conceived the idea of using a tail rotor to counter the torque effect on single-rotor helicopters. But torque effect was only one problem which had to be overcome before fully successful helicopters appeared. Cornu's helicopter and the Breguet-Richet *Gyroplane No. 1* flew on several occasions; the latter machine during one test unexpectedly gained sufficient momentum to finish damaged in a field of vegetables. Major new problems with the stability of helicopters showed themselves only after horizontal flight was attempted, but mostly in the case of single-rotor helicopters.

The fully developed helicopter incorporates several innovations to its rotor system that were hard learned by the pioneers of rotary-wing craft. When a helicopter is in vertical or hovering flight, the amount of *lift* created by each rotor blade remains equal and the helicopter is stable. However, when horizontal flight is attempted, new problems arise that especially

The American English helicopter, which broke loose during tethered trials.

effect single-rotor craft. When a helicopter moves in any direction, the lift becomes unequal from the turning rotor. This is simple to explain. Like the wing of a conventional aeroplane, each rotor blade creates lift in proportion to the airflow over its surface. Blades turning in rotation, therefore, develop more lift as they advance in the same direction as the helicopter travels, as the helicopter's own speed is added to the constant speed of rotation. As blades retreat away from the helicopter's direction of movement, the helicopter's own speed of travel is subtracted from the speed of rotation. Therefore, one side of the rotor creates more lift than the other, despite a constant rotational speed. This effect on single-rotor machines was well demonstrated by Cierva Autogiros of the early 1920s, which sometimes literally overturned. The effect was overcome by the adoption of the flapping hinge. Instead of each blade being attached rigidly to the rotor head, each is joined to the rotor head via a flapping hinge. This allows each blade to travel marginally up and down. Therefore, as each blade advances it rises on the flapping hinge as increased lift is created, while each retreating blade falls as lift decreases.

Another effect on the rotor is that advancing blades try to gather speed, while retreating blades try to slow, again despite a constant rotational speed. Developed helicopters have what are known as drag hinges with dampers. The drag hinges give the blades limited horizontal movement to allow for the effect of acceleration while advancing, while the dampers ensure that this movement is kept to a minimum. Without the drag hinges the blades could distort. Without dampers the rotor's centre of gravity could be altered, resulting in dangerous vibration.

A further movement of the rotor blades is effected by the pilot, who can adjust the pitch (angle about lateral axis) of the blades using two separate levers in his cockpit. Using the collective pitch lever or 'stick', the pilot can simultaneously change the pitch of all the blades for vertical flight control. To achieve directional control it is necessary to tilt the rotor in the direction in which the helicopter is to travel: forward, backward or sideways. Using the cyclic pitch control stick, the pilot can affect this tilt. Basically, this control

American Luyties Otto helicopter that appeared in 1908 and is said to have flown on several occasions.

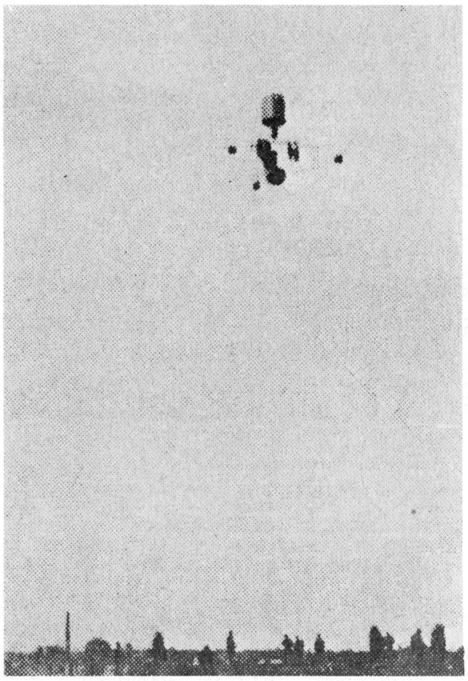

Right: This poor quality photograph is nevertheless of great interest, showing Oberstleutnant Stefan von Petroczy's remarkable Le-Rhône-engined observation helicopter in full tethered flight during the First World War.

Above: Ellehammer's helicopter on its first flight in 1912, with Jacob Ellehammer standing by the hovering machine about to switch off the ignition. He autographed the photograph two years later.

alters the pitch of each blade continuously during each rotational cycle, via the swashplate, increasing the pitch of blades moving away from the helicopter's intended direction of travel while decreasing the pitch of those blades in the advancing stage. As increased pitch forces blades to rise, a tilt effect is achieved and the rotor provides lift and propulsion.

As mentioned previously, most developed helicopters use a tail rotor to counter the torque effect and to keep the fuselage straight. By varying the thrust of the tail rotor, the fuselage can be made to swing round. However, most early helicopters adopted contra-rotating rotors, which automatically countered torque and so removed the need for a tail rotor. These were more stable in horizontal movement than single-rotor helicopters until the invention of flapping hinges.

A Danish aviation pioneer by the name of Jacob C. H. Ellehammer, who produced unusually configured aeroplanes which 'hopped' and then flew from 1906, is also credited with a helicopter that lifted itself and the pilot in 1912 and featured a simple form of cyclic pitch control. However, six years before this, G. A. Crocco had suggested the necessity of cyclic pitch control. The Ellehammer helicopter had co-axial contra-rotating rotors powered by a single radial engine of 36 hp. Each rotor had a diameter of 24 feet 6 inches (7.47 metres); the four wire-braced blades of the upper rotor were attached to an open circular structure while the lower blades were attached to a covered structure.

Little else of importance occurred until 1915. Breaking new ground after the experiments of W. H. Phillips in 1842, Frenchmen Papin and Rouilly built a helicopter whose single hollow rotor blade was driven by air produced from an 80-hp Le Rhône engine-powered fan. The air had two possible outlets, one through the rotor and another through a pipe intended to provide some thrust for controlling the direction of flight. The helicopter was possibly the first to be tested from water, undoubtedly in the mistaken belief that a crash would be less severe on water. On 31 March 1915 the helicopter, resting on Lake Cercey (not far from Dijon), was started but the rotor made it oscillate badly and the pilot jumped off in time to see the machine sink.

Remarkably little progress in the development of the helicopter was achieved by the major powers during the First World War, despite the realization many years before hostilities broke out that the helicopter could offer an air observation platform that was less of a target than a balloon or kite and which did not continually have to fly in circles as did the observation aeroplane. However, often overlooked are the wartime experiments of Oberstleutnant Stefan von Petroczy of the Austrian Army Balloon Corps, and Dr. Ing Theodor von Kármán. These experiments were aimed at producing a tethered observation helicopter with defensive armament.

Having conducted experiments on several model helicopters, first using twisted rubber as the power source and then a 5-hp compressed air motor, two full-size helicopters were completed for tethered trials. The first was powered by a light but powerful electric motor, which it was hoped would generate 225 hp but only managed 190 hp. Nevertheless, with this motor a crew of three was lifted. However, after running for about 30 minutes the motor burned out. The second von Petroczy helicopter was somewhat different in conception, having a triple-arm structure constructed from steel tubing, each arm carrying a 120-hp Le Rhône engine. The power from all three was used to drive contra-rotating rotors of about 19 feet 8 inches (6 metres) in diameter. Below the main structure were four inflated balls that acted as the landing gear and shock absorbers. Perhaps the most interesting feature of the helicopter was the cylindrical cockpit provided above the rotors, designed to accommodate an observer and gunner and supported on a strut that passed through the hollow rotor shafts. The helicopter was designed to be winched out to height with the rotors turning and winched in after the observation flight. Attack by an enemy could be dealt with in two ways, by the gun or by a stowed parachute which was large enough to support the complete helicopter for retrieval with engines stopped. Interestingly the

Louis Brennan standing by his helicopter at Farnborough. Note the engine-driven rotor-tip propellers.

parachute was designed to be deployed mechanically by the action of the crew or by a device that was triggered to operate when and if the revolution of the rotors fell below a safe rate.

During trials the helicopter demonstrated its ability to stay aloft for an hour at low level, while subsequent tests saw an altitude of 160 feet (50 metres) being attained. It remained stable in wind conditions of 20 mph (32 km/h) and lifted a weight equivalent to four men. Although after 14 test flights the helicopter's engines were in need of attention, the helicopter unfortunately was ordered up once again for official inspection. During this flight the engines failed to run smoothly and the helicopter became unstable. Those on board jumped out of the cockpit and returned to the ground by parachute, while the winch attempted to retrieve the helicopter. On landing, however, the machine fell onto its side and the rotors were wrecked.

In March 1918 von Kárman, with the assistance of Wilhelm Zurovec, completed a new helicopter in Budapest known as the PKZ 1. Like the original full-size von Petroczy helicopter, it adopted an electric motor as its power source, although on this occasion it drove four rotors. It is believed that four flights were made in tethered form and on three of these occasions it carried a crew of three. Wilhelm Zurovec then went on to demonstrate his own helicopter on 2 April the same year, with the same engine/rotor layout as von Petroczy's second full-size helicopter. It achieved a similar tethered performance. It is worthy of mention that von Kárman joined the California Institute of Technology in 1926, where he remained until 1949. During this period he started the first rocket motor project on behalf of the U.S. Army and in 1944 also took up the position of chairman of the U.S.A.A.F's Scientific Advisory Board. He later became one of the foremost scientists working in the field of space exploration. He eventually died in 1963.

Realizing limitations

For a few years from 1918 it appeared likely that development of satisfactory helicopters was just around the corner. Yet the likely limitations of any helicopter of the period were recognized: a postwar proposal that the British Air Ministry should allocate a prize of £50,000 for a satisfactory helicopter only suggested a performance equal to that of a 1910 aeroplane. Of course even this performance was well beyond the capability of any existing helicopter and most of the 'interwar' period was to pass before a helicopter could demonstrate worthwhile performance. This is not to say there was no progress. The opposite is the case. At Farnborough in Britain, work on a helicopter designed by Louis Brennan, the inventor of the Brennan torpedo, was undertaken in great secrecy and it was reported that manned flights were achieved. In the United States of America, the helicopters built from 1918 that flew included the U.S. Army-financed De Bothezat and 80-hp Berliner helicopters and the 126-hp Crocker Hewitt helicopter, with contra-rotating rotors. The Berliner helicopter, the result of experiments begun in 1908, used the same form of downwash control as that adopted by Cornu in 1907. But the real progress was made, as before the war, in France.

A name prominent in the annals of aviation history is that of the Argentinian-born Marquis de Pateras Pescara. It appears that his first aircraft design was for a small seaplane, a model of which was flown in 1912. From 1919 to 1925 Pescara produced a series of full-size helicopters. His No. 3 version has been credited with first demonstrating successful cyclic pitch control via a tilting rotor head. His first model was a 45-hp 24-blade

Left: Pescara 4S in France, one of the first Pescara helicopters to demonstrate successful cyclic pitch control.

Below: Oehmichen No 2, first flown on 11 November 1922. On 14 April 1924 it managed a straight line distance of 1,181 ft (360 m), extended to 1,722 ft (525 m) on the 17th and a greater distance the following day. Its time over the 1-km closed circuit on 4 May was 7 minutes 40 seconds. Notwithstanding these records, it was an impractical helicopter.

helicopter, built in Barcelona, Spain, but this failed to fly. Pescara's second helicopter, with a 170-hp Le Rhône engine, lifted in May 1921. His research drew the attention of the French authorities and further work was conducted in that country. The Pescara No. 3 proved capable of lifting the pilot, albeit in somewhat directionally unstable flight. This machine used a co-axial contra-rotating rotor layout, each rotor comprising a cruciform of biplane surfaces. Warping was employed to alter the pitch of the blades. The engine (a 180-hp Hispano-Suiza) and the pilot's seat and controls were carried in a strong structure beneath the rotors, supported on wheels. Apart from having the ability to select collective pitch and cyclic pitch control, the pilot could also reverse the pitch of the rotor blades so that, if engine failure occurred in flight, the rotors would turn in autorotation to allow a safe 'power off' descent and landing.

While Pescara is credited with the first helicopter to demonstrate successful cyclic pitch control, he was not alone in its application at that time. Frenchman M. Damblanc constructed an experimental helicopter in 1920 to the order and expense of the French government. This machine, known as the *Alerion*, had the wings, fuselage and tail unit of an aeroplane (monoplane) but with two coupled engines that powered co-axial contra-rotating four-blade rotors. Like Pescara's helicopters, this was flight tested with cyclic pitch control and the rotors could be autorotated if required. It is known that trials on this machine ended in 1922, or thereabouts, and M. Damblanc thereafter began new experiments.

To Pescara helicopter No. 3 went a world record on 18 April 1924, when it managed a distance of 2,415 feet (736 metres) at Issy-les-Moulineaux. On 4 May the same year, the French Oehmichen No. 2 demonstrated for the first time a helicopter flight of one kilometre in a closed circuit, but this was a totally impractical machine. Based on a cruciform structure, the single 180-hp Gnome engine drove four rotors for lift and eight small vertical propellers for directional control. Two different-size rotors were adopted, two of 24 feet 11 inches (7.6 metres) and two of 21 feet (6.4 metres).

One example of a modern helicopter with co-axially mounted contra-rotating rotors is the Sikorsky S-69, seen here with auxiliary turbojet engines carried in pods on the fuselage sides. Intended for research into the Advanced Blade Concept (ABC) rotor system, it first flew in 1973.

The Pescara helicopter can be said to have peaked early helicopter development, an era which had stretched over centuries. From a practical point of view, the wide adoption of the twin equal-size rotor layout had been sensible given the 'state of the art', countering as it did the problems associated with torque effect and major instability in horizontal flight. The penalty was added complications to the mechanical systems and the resulting heavy weight.

Most fully practical helicopters that appeared in the 1940s were single-rotor machines unless multirotor by virtue of large size. There were, of course, exceptions to this and even in the 1980s co-axial rotor helicopters of modest size are flown. The credit for developing the practical single-rotor craft is given to the Spaniard, Juan de la Cierva, whose autogyros (and those of others) filled the time gap between pioneer helicopters and the appearance of practical helicopters.

Propping the Rotor

The period between 1920 and 1923 was important in the development of rotary-winged aircraft for many reasons, as outlined in the first chapter. As we have seen, the Damblanc helicopter and those constructed by Pescara were designed to allow for autorotation in case of engine failure, thus greatly improving safety. However, by far the most important experiments in autorotation were being conducted in Spain during this period, and these eventually led to the first rotary-winged aircraft to go into production for both commercial and military service.

Don Juan de la Cierva was born in Spain in 1886. Soon after the end of the First World War he began work on a new form of rotorcraft, of a type related to the helicopter but with important differences. Cierva's intention was to construct an 'autogyro', which is basically the term for a hybrid aeroplane and helicopter. Realizing that the forward motion of an aircraft alone could spin a freely mounted rotor to a point at which sufficient lift could be gained from the rotor for flight, thus obviating the need to power the rotor, he set to work constructing his C.1. This machine was a 1911 Deperdussin tractor-engined monoplane with its wings removed and co-axial freely turning rotors mounted to the rear of the pilot's cockpit. Appearing in 1920, this model failed to fly. His next two *Autogiros* (as Cierva called them) were also unsuccessful, but this time because of the unstabilizing effect of rigidly attached rotor blades on a single rotor machine (see previous chapter). Then came the breakthrough.

In 1922 Cierva built his C.4 Autogiro. This machine once again used a conventional aeroplane fuselage and tractor engine, but this time ailerons were carried on outriggers where the lower wing of an aeroplane would have been and the four blades of the single rotor were attached to the head by flapping hinges. Why flapping hinges were necessary for stable flight is fully covered in the opening chapter and it is sufficient to say here that the C.4 became the first-ever successful gyroplane. It made its maiden flight on 9 January 1923. To Cierva is credited the innovation of flapping hinges and the C.4 was certainly the first free-flying rotary-winged aircraft of any practical use.

From the C.4 Cierva developed the C.5 and then, with financial support from the Spanish government, the superior C.6A. This Autogiro was based upon an Avro 504K and was thus powered by a 110-hp Le Rhône engine. In general configuration the C.6A was similar to the C.4, with the articulated rotor pylon-mounted ahead of the pilot's cockpit, but it adopted a much wider main undercarriage for

Don Juan de la Cierva's unsuccessful first Autogiro, *the C.1.*

increased stability on the ground. Assembled in Madrid, it was first flown by a military pilot in May 1924. During the remainder of that year the C.6A underwent trials, during which an altitude of approximately 655 feet (200 metres) was attained and minimum and maximum speeds were found to be 16 mph (25 km/h) and 68 mph (110 km/h) respectively. The crowning achievement of the year was a flight of 7½ miles (12 km) from the test ground to Getafe. Similar to the C.6A Autogiro was the C.6B.

On 15 October 1925 the C.6A was flown at Farnborough, England, for an invited audience and eventually the Air Ministry ordered two examples from Avro as the C.6C single-seater and C.6D two-seater. The C.6C, built under licence from Cierva and powered by a 130-hp Clerget engine, was the first to appear, in mid-1926. This seems to have been the only one of the two to be allocated a military serial and is remembered as the R.A.F's first rotary-winged aircraft. Of course this machine was built for test purposes only and so never joined an active squadron, making the R.A.F. connection somewhat academic. Meanwhile the C.6D, which had appeared in July 1926, became the world's first two-seat autogyro and on 30 July Cierva himself became the first-ever passenger in the machine.

Left: Cierva C.4; the four blades are attached to the rotor head by flapping hinges.

Below: Co-operation between the Cierva Autogiro Company and Westland Aircraft resulted in the Westland C.29. Intended to accommodate the pilot and four passengers, it proved to possess major ground resonance problems and was abandoned in 1936 without a flight

Even before the Avro-built C.6s flew, the Cierva Autogiro Company had been founded in Britain by Commodore J. G. Weir (24 March 1926) to exploit the designs of Cierva. Indeed, over the following years Cierva licenced Autogiro production to a number of companies, although A. V. Roe (Avro) produced by far the largest number. Other British companies that built Cierva craft of many differing models included the British Aircraft Manufacturing Company, de Havilland Aircraft and Westland Aircraft. Of the C.8 models built by Avro and flown for the first time between 1927 and 1929, the C.8L Mk II went to Weir, the Mk III was sold to the Italian government and the MK IV was delivered to Harold F. Pitcairn in the U.S.A. The Mk II, civil registered *G-EBYY*, is remembered as the first-ever rotary-winged aircraft to fly across the English Channel when, on 18 September 1928, Cierva left Croydon and flew himself and a passenger to Le Bourget. It is of interest to note that Cierva had only learned to fly in 1927 at a British aero club. The C.8L Mk II was powered by a 200-hp Armstrong Siddeley Lynx IVc engine, which bestowed a maximum speed of 100 mph (161 km/h). A maximum range of 255 miles (410 km) was possible.

The C.8L Mk IV delivered to Pitcairn was also of interest. Harold Pitcairn had been the founder of an aeroplane manufacturing company bearing his name and based at Willow Grove, Philadelphia. He received a licence to build the Cierva Autogiro in the U.S.A. and his company was subsequently renamed the Pitcairn Autogiro Company. More important to U.S. military aviation, Cierva and Pitcairn licenced the Kellett Autogiro Corporation of Philadelphia to build Autogiros and the developed Kellett YG-1 of the late 1930s became the U.S. Army's first rotary-winged aircraft. Meanwhile 12 Avro Rota Mk ls, Cierva C.30As licence-built on a non-exclusive basis, had been delivered to the R.A.F. between 1934 and 1935 as that force's first rotary-winged aircraft in service use. These, each powered by a 140-hp Siddeley Genet Major I.A. engine and possessing a performance similar to that of the C.8L Mk II, initially joined the Army co-operation school at Old Sarum. Avro

On exhibition in France, the Cierva C.8L Mk II G-EBYY *was the first rotary-winged aircraft to fly across the English Channel.*

Avro-built Cierva C.30A.

built 78 autogyros of C.30A type alone, most of which were civil registered for the British market and for export, although military users in addition to the R.A.F. included the air forces of Belgium and Yugoslavia. Cierva C.30As were also built under licence in France and Germany. The French company, Lioré-et-Olivier, built 25 with 175-hp Salmson engines for the Armée de l'Air and 40 others came from Focke-Wulf in Germany under a September 1931 agreement.

The Cierva Autogiros are undoubtedly the best remembered autogyros of the interwar period and were flown in many countries. A trick learned back in the mid-1920s, to pre-spin the rotor manually to reduce the already short take-off distance, was later taken a stage further when a subsequent C.30 model used a gearing arrangement that allowed the engine to spin the rotor before take-off but to disengage before flight. This technique, known as jump start, has recently been revived for small autogyros available to enthusiasts today (described later in this chapter). By using the engine to pre-spin the rotor, the Autogiro came even closer to the helicopter but, while it could fly very slowly indeed and land without forward roll, it still lacked the helicopter's ability to take off vertically and to hover. After the end of the Second World War the re-formed Cierva company went into helicopter manufacture.

Pitcairn in the U.S.A. produced several models of autogyro; its open-cockpit PA-18 two-seater and PA-19 four-seat cabin types with 160-hp Kinner R-5 and 420-hp Wright R-975E-2 Whirlwind engines respectively sold reasonably well. Kellett autogyros were historically more important. It was a Kellett K-3 that went on the 1933

Antarctic expedition led by Admiral Byrd. The KD-1 made the first-ever aircraft landing on the roof of a building, during an experimental airmail service to mark the opening of the Philadelphia post office in May 1935, and the KD-1B, in the livery of Eastern Air Lines, flew the first-ever scheduled airmail service by rotorcraft between Camden Airport and the roof of the Philadelphia post office, from 6 July 1939. The KD-1B was a single-seater capable of 127 mph (204 km/h), with the pilot protected from the elements under a fighter-type sliding canopy. Like the Cierva company, Kellett turned to helicopters during the Second World War and its XH-8 for the U.S.A.A.F. flew in August 1944.

Other nations in which early helicopter development temporarily gave way to the autogyro included the Soviet Union. The first successful Soviet rotary-winged aircraft was the KaSkr-1, the work of Nikolai I. Kamov and Nikolai Kirillovich Skrzhinskii. This autogyro was based on the fuselage of a U-1 aeroplane and began flight trials in September 1929. Unfortunately it proved to be underpowered and went back to the factory for modification. It re-emerged in the following year as the KaSkr-2 with a Soviet licence-built 240-hp Gnome-Rhône Titan replacing the original 120-hp Le Rhône (Soviet M-2) engine. Kamov continued working with autogyros until 1943, when he switched to helicopters. Kamov helicopters subsequently became standard Soviet types for military and civil use.

Assisting with the design of the KaSkr-1 was a man by the name of Mikhail Leontyevich Mil who, after the Second World War, also designed helicopters for widespread use. However, the most important work on autogyros in the Soviet Union was achieved at the Tsentralnyi Aero-gidrodinamicheskii Institut, the Central Aero-Hydrodynamics Institute (ANT or TsAGI) in Moscow, a scientific institute founded in 1918, at which Kamov,

One of five Rota Mk 2s (Cierva C.40s) ordered by the Air Ministry to specification 2/36, and built by Avro as follow-ups to the Rota Mk 1s.

Skrzhinskii and Mil worked from 1931. The first TsAGI autogyro was the 2-EA, a Titan-engined two-seater of 1931 that possessed a maximum speed of nearly 100 mph (160 km/h) and an endurance of well over an hour. It incorporated in its design a twin-fin and rudder tail unit and narrow-chord, strut-braced, low-mounted dihedral wings. It was used in an extensive development programme which lasted two years.

The next important autogyro from the TsAGI was the A-4, which has been attributed to both Nikolai Dmitrievich Kuznetsov and Skrzhinskii. It was basically a more developed production-suitable version of the 2-EA, powered by a Soviet licence-built version of the Titan (known as the M-26 and developing 300 hp) and featuring a ring cowling for the engine and a new tail unit with a deep fin and unbalanced rudder. Capable of 109 mph (176 km/h), a very small number joined Soviet forces as rotorcraft trainers from 1934.

While work on the A-4 progressed, Kamov simultaneously headed the development of the A-7. The original A-7 appeared in September 1934 as a two-seater with a large 480-hp M-22 engine (a licence-built Bristol Jupiter). Two improved variants became the A-7*bis* and A-7Za. The A-7 was not only the first Soviet autogyro to feature pre-flight rotor spin but was the first in the world to be armed with machine guns. A number of production examples were produced for squadron service as reconnaissance and spotting aircraft and A-7Zas carried three 7.62-mm PV-1 machine guns in forward-firing and rear defence positions; Mikhail Mil served with an A-7 unit during the German invasion of the Soviet Union from 1941, supporting the Soviet Army. A-7s had previously been used for forestry and exploration work in Tien Shan.

The second Soviet autogyro to fly without the familiar auxiliary wings was the TsAGI's A-14 of 1935 (the first had been the single-seat

Below: Eastern Air Lines operated Kellett KD-1B, used on the world's first scheduled air mail service by rotorcraft.

Right: Soviet TsAGI A-4 autogyro.

Captured example of the Focke Achgelis Fa 330 Bachstelze submarine rotor-kite, on exhibition in London.

A-12, fitted with a 640-hp Wright Cyclone F-3 engine), but this very low-powered machine was intended only for experimental use. The final autogyro from the TsAGI was the A-15, an armed two-seater of 1937 with the most powerful engine fitted to any aircraft of this type. Remaining a prototype and possibly not even flown, it was installed with a 730-hp M-25V development of the U.S. Wright Cyclone.

During the Second World War autogyros were employed on a minor scale by the forces of Britain, the U.S.A., Germany and the Soviet Union. Many civil machines were impressed for military service. The roles to which these autogyros were put were generally non-combat, although it is entirely possible that Soviet A-7s might have been the first autogyros to have opened fire on an enemy. However, in Japan the Kellett KC-1A was produced by Kayaba as the Ka-1, and entered Army and Navy service on a considerable scale for observation (Army) and anti-submarine patrol (Navy). Ka-ls could carry two light bombs or depth charges and operated from shore and aircraft carriers. It is also interesting to note that the German Navy had expected to be able to deploy autogyros from their ships before the war, but the Flettner Fl 184 prototype of 1935 was destroyed by fire before evaluation and so the design was dropped.

Development of more versatile helicopters overtook that of autogyros during the wartime period, virtually without exception. Germany, however, deployed the Focke-Achgelis Fa 330 Bachstelze during the war, but it was not really an autogyro in the accepted sense but rather a rotor-kite. A diminutive single-seater, it was designed to be stowed on board German IX-type U-boats in dismantled form. The intention of carrying the Fa 330 was simple enough. By towing the rotor-kite by cable behind and above the surfaced submarine, its pilot could search for enemy shipping 'over the horizon'. The Fa 330 could attain an altitude of about 400 feet (120 metres) but required a minimum airspeed of 17 mph (27 km/h) to stay up.

Observations by Fa 330 pilots were relayed to the towing submarines by telephone. However, while the concept was sound, in service some submarine crews looked upon the Fa 330 as a potential hazard to their safety. If the submarine was spotted while surfaced with the Fa 330 deployed, two courses of action lay open. The first was to winch-in the rotor-kite at speed before crash diving, while the second relied on the pilot jettisoning both the rotor and towing cable. In the latter action, the released rotor automatically deployed the pilot's parachute. Whether the submarine's crew waited for the pilot to return to ship or decided to pick him up later, the delay in crash-diving could spell trouble. In the event, it is considered that about 200 Fa 330s were built, going into service from 1942. Fa 330 operations in the Atlantic were deemed very risky and so the Fa 330's best work was performed in other theatres of war.

By the cessation of hostilities in 1945, the helicopter had taken its first tentative steps into military service and civil operations were around the corner. The days of the sophisticated autogyro were just about over and few appeared to mourn their passing. Yet in many respects the heyday of the autogyro was still to come. During the 1950s Igor Bensen, formerly employed by the Kaman Aircraft Corporation, developed a series of towed rotor-kites and autogyros for sale to private individuals either as completed aircraft or in kit form, or plans could be purchased by anyone with the ability to construct the craft from bought materials. The hallmark of Bensen rotorcraft was simplicity.

Bensen offered the unpowered and powered models with wheel and float undercarriages. One rotor-kite model even sported a dinghy hull which made it the B-8B Gyro-Boat, the first rotorcraft to be literally a 'flying-boat'. Bensen autogyros became extremely popular all over the world. The initial powered version was the B-7M Gyro-Copter, first flown on 6

December 1955. This had a very basic fuselage structure of aluminium tubing, onto which was fixed the single seat for the pilot. Above the pilot's head the two-blade rotor was carried and to his rear the 40-hp Nelson H-59 pusher engine and propeller used to achieve forward motion. A vertical fin and rudder were employed. Although unable to take off vertically and hover, the Gyro-Copter could take off in short distances. This set the pattern for all subsequent Bensen powered autogyros.

Produced in the greatest numbers

More than a quarter of a century later the Bensen Gyro-Copter is still going strong, together with its twin-float counterpart, the Hydro-Copter, and the Gyro-Glider. By 1984 about 10,000 sets of plans alone had been sold worldwide to enthusiasts wishing to construct their own Bensen machines. Today's range of Bensen aircraft centres on the B-8M and B-8V Gyro-Copters which offer more power and better performance than the original B-7M. The B-8M adopts a 72-hp McCulloch two-stroke engine for forward power, while the 'V' uses a modified Volkswagen automobile engine. In either case a mechanical drive system for the rotor can be incorporated, which allows the engine to spin the rotor prior to ground roll. Once the rotor is up to flying speed, the engine power is returned to its normal pusher propeller and the rotor autorotates to allow a lift-off in a distance of just 50 feet (15 metres).

Recent developments by Bensen have brought his diminutive machines even closer to helicopters. In the 1970s the Model B-8MJ was developed, which was a B-8M with a Power Head to allow a jump take-off without any ground roll. The B-8MJ was still not capable of hovering and so Bensen developed simultaneously the B-18 Hover-Gyro, which marries the autogyro and helicopter. On this aircraft co-axial rotors are used, the upper rotor autorotating and the lower powered by a second engine. Going a stage further, Bensen demonstrated its Super Gyro-Copter in 1981, in which a small amount of the single pusher engine's power is bled to drive the single rotor continuously.

Bensen Aircraft is just one of a number of companies making autogyros available for amateur construction by enthusiasts, although no other firm has matched sales. Bensen autogyros have provided inexpensive forms of rotorcraft to the public and yet their simplicity has not meant poor performance. Indeed, a B-8M can fly at 85 mph (137 km/h), has a service ceiling of 12,500 feet (3,810 metres) and its normal range on 6 U.S. gallons (22¾ litres) of fuel is 100 miles (160 km).

Those followers of James Bond, Ian Fleming's fictitious British agent, have probably seen one of the world's finest autogyros in action on the movie screen without realizing it. During the Bond film, *You Only Live Twice*, a missile-armed autogyro cleverly disposed of the enemy during a dramatic air-to-air sequence. The autogyro was *Little Nellie*, just one of a fleet designed, built and flown by Wing Commander K.H. Wallis. Constructed in England, Wallis autogyros currently hold the world height record at 18,516 feet (5,644 metres), distance in a straight line record at 543.274 miles (874.315 km), distance in a closed circuit record at 416.48 miles (670.26 km), and speed in a straight line record at 111.225 mph (179 km/h) in the FAI's Class E.3.

Apart from high performance, Wallis aircraft have demonstrated practical applications and were even tested by the British Army for certain military tasks. They have been used as aerial photographic platforms, for the detection of illicit graves in co-operation with the Home Office and police, aerial colour stereo-photographic survey, and other roles which have included a part in the search for the Loch Ness Monster in Scotland. The first Wallis autogyro appeared and flew in 1961 and, while production nearly got underway many years

Left: Bensen B-8M Gyro-Copter, a light autogyro that has been assembled in vast numbers since first appearing in 1957.

Above: Wallis WA-116/Mc G-ATHM. *Built in 1965 and initially flown on a tea plantation in Sri Lanka, it was converted in 1974 for record breaking and in July that year set the first speed records for autogyros in a 500-km closed circuit.*

ago, it is only now that production is planned for military and commercial applications in association with W. Vinten Limited, a company specializing in aerial reconnaissance equipment. Possible uses in the commercial field will be photogaphy, geophysical survey, pipeline inspection and crop spraying. The production machine will be the WA-116/W, a 75-hp Weslake-engined single-seater capable of cruising at 115 mph (185 km/h) and achieving a still-air range with normal fuel of 300 miles (483 km); its endurance will be up to a maximum of 4 hours. Unlike the American Bensen craft mentioned earlier, Wallis autogyros have streamlined cockpit fairings.

Convertiplanes

The Bensen Aircraft Corporation is by no means the first to combine the attributes of the helicopter and autogyro to produce a rotary-winged aircraft with the economy of operation of an autogyro and the vertical and hovering flight of a helicopter. Back in the 1950s and early 1960s huge aircraft were developed along this line, with the addition of short-span wings to make possible high-speed horizontal cruising flight.

Representing only one form of this class of aircraft, known as the convertiplane, probably the best remembered exponent was the British Fairey Rotodyne. The story of the Rotodyne really begins in December 1947, when Fairey flew, for the first time, a fairly small four-/five-seat experimental helicopter known as the Gyrodyne. This incorporated several innovations, the most obvious of which was the deletion of an anti-torque rotor. Instead, the 525-hp Alvis Leonides radial piston engine powered both the single main rotor and a conventional tractor propeller carried at the tip of the starboard stub-wing. With this arrangement the Gyrodyne could be flown either as a helicopter or as an autogyro, and on 28 June 1948 it set a new world speed record for

Opposite: Fairey Rotodyne Y at Farnborough, which so nearly became the world's first large commercially operated convertiplane.

helicopters at 124.3 mph (200 km/h), the first time 200km/h had been reached by a rotary-winged aircraft. (However, it would be incorrect to believe that the Gyrodyne was the first aircraft capable of autorotation to employ a sizable passenger cabin, as the prewar Pitcairn PA-19 Autogiro, to name but one, could accommodate four persons while flying on the power of a 420hp Wright Whirlwind radial.)

After the Gyrodyne crashed in April 1949, just days before it was due to make an attempt on the 100-km closed-circuit speed record for helicopters, it was rebuilt as the Jet Gyrodyne. The rotor was now turned by tip-jets and the smaller vertical propeller reversed to pusher configuration. The construction and flight-testing of the Gyrodyne/Jet Gyrodyne assisted in the design of Fairey's incredible Rotodyne, which was expected to become the world's first large passenger-carrying convertiplane.

The Rotodyne was to break new ground in the field of rotorcraft. Fairey's initial model was the 40-passenger Rotodyne Y and the company intended the 54/70-passenger Rotodyne Z to become the production version. Making its maiden flight on 6 November 1957, the Rotodyne Y was powered by two 3,000-ehp Napier Eland N.E.L.3 turboprop engines carried in nacelles under the 46-foot 6-inch (14.17-metre) wings and driving four-blade tractor propellers. The single four-blade rotor was carried on a pylon high above the square-section monocoque fuselage, on the rear of which was carried a twin fin and rudder tail unit.

The principle of the layout closely followed the earlier experiments with the Gyrodynes. At the tip of each rotor blade was carried a tip-jet. Air compressors, powered by the main engines, delivered air to the tip-jets, where fuel was burned to produce thrust for rotation. In this manner the Rotodyne could operate as a helicopter, with the propellers producing no forward thrust. For horizontal cruising flight the pilot could select a new mode, transferring all engine power to the tractor propellers and allowing the rotor to turn freely in autorotation. Combined with the wings, the rotor was sufficient to maintain lift while the engines gave a maximum speed in level flight of 185 mph (298 km/h), far in excess of conventional helicopters. In actual operation, the pilot would select the helicopter mode for take-off and landing and autorotation for cruising over a distance of up to 450 miles (724 km).

The all-important first transition from vertical flight to horizontal cruising was demonstrated on 10 April 1958. On 5 January 1959 the Rotodyne Y was flown at a speed of 190.9 mph (307.22 km/h) over a 100-km closed-circuit, setting a world record that remains good today. One problem that beset the development programme was excessive noise from the tip-jets. Considerable time and expense went into the development of suppressors to reduce this but throughout the life of the Rotodyne it remained a problem.

The production Rotodyne Z was expected to adopt suppressors that could reduce the aircraft's noise level to about 95 decibels at a distance of 200 feet (61 metres), despite its much more powerful 5,250-shp Rolls-Royce Tyne turboprop engines and greater overall size. A number of helicopter-operating companies became extremely interested in the Rotodyne and early orders were placed by Okanagan Helicopters of

Below: Fairey Gyrodyne at White Waltham in June 1948.

Canada and New York Airways from the U.S.A., while BEA was negotiating the purchase of six in 1959. However, despite its incredible potential as a commercial and military transport (rear clamshell loading doors allowed the carriage of passengers, freight and/or vehicles), the complete programme was cancelled in 1962 owing to the escalating cost of developing tip-jet suppressors and other factors. It is of interest to note that Fairey believed the Rotodyne Z would have been capable of cruising at 201 mph (323 km/h) and cover average stage lengths of more than 250 miles (402 km) with 57 passengers on board. Maximum range might have been some 650 miles (1,046 km), a long range for rotorcraft.

Another convertiplane of the late 1950s was the Soviet Kamov Ka-22 Vintokryl, of a size similar to that of the Rotodyne Z but of very different configuration. This was almost certainly designed to be compatible with the Antonov An-12 turboprop transport aeroplane then under development, thus allowing bulk freight, 100 troops/paratroops, vehicles or civil passengers to be carried into less accessible areas. Far less attractive than the Rotodyne and without the British aircraft's retractable undercarriage, the Vintokryl nevertheless established a world speed-in-a-straight-line record of 221.4 mph (356.3 km/h) on 7 October 1961 and a height record, carrying a 15,000 kg payload, 8,491 feet (2,588

Kamov Ka-22 Vintokryl in the skies over Tushino in July 1961.

metres) on 24 November the same year, which have never been bettered.

In operation, the Vintokryl flew on the same principles as the Rotodyne, although the twin main rotors, mounted above the 5,622-ehp Soloviev TV-2 turboshaft engines at the wingtips, were driven directly by the engines when flying in helicopter mode. Thus it can be assumed that the Vintokryl did not suffer from excessive noise in the manner of the British convertiplane. Nothing was known of this aircraft outside the Soviet Union until it was seen during the Soviet Aviation Day display at Tushino in July 1961, creating more than a little surprise for the Western onlookers. It was then correctly described by the Soviet commentator as the most powerful vertical take-off aircraft in the world but nothing further was heard of it and so it joined the Rotodyne as merely a milestone in the annals of aviation history.

The concept of the convertiplane took on many forms during the 1950s, some of which are described later in this book. The idea of a transport aircraft capable of vertical operation into and out of difficult areas and yet possessing performance more closely associated with fixed-wing aircraft has never been laid to rest and even today experiments in this field continue. Perhaps the most surprising aspect of all postwar research is that not one transport convertiplane has ever gone into commercial or military service proper, leaving vertical operations entirely in the hands of the helicopter.

Threshold of Success

The lack of substantial progress with helicopters by the early 1920s and the emergence of the practical autogyro did not deter all further helicopter experiments. Indeed, it was often as not a shortage of funds that prevented individuals from progressing their work.

One rather fascinating fact is that some early Russian helicopter experiments are inextricably and historically bound to the development of the first successful American helicopter, by the efforts of a man who was certainly not short of finance for his early work but who gave up helicopters after early failures to concentrate on aeroplanes. In Kiev in 1889 Igor Ivanovitch Sikorsky was born. His parents were both doctors and at a very early age his attention was turned to aviation through their keen interest in the work of the Italian, Leonardo da Vinci. At the age of 12 Sikorsky built and flew his first aircraft – a tiny model helicopter powered by twisted rubber.

At the age of about 19 Sikorsky travelled to France, then considered the mecca of European helicopter and aeroplane development, and returned to produce his first full-size helicopter with a 25-hp Anzani engine he had purchased. Typically, his first helicopter was of the co-axial contra-rotating type, and it failed completely when tested in 1909. A second similar machine, built in the following year, faired little better and only managed to rise a short distance above the ground and then without Sikorsky on board. Such poor results were enough to turn his attention away from helicopters temporarily, although his subsequent outstanding success with aeroplanes did not block his mind from rotorcraft entirely.

Above: Igor Sikorsky's first attempt at a full-size helicopter, tested in 1909 but a failure.

Left: Sikorsky's second helicopter, built in 1910. Mounted on rails, it just managed to lift from the ground but was insufficiently successful to prevent Sikorsky from taking up the design of fixed-wing aircraft.

Encouraged by his parents, Sikorsky built an aeroplane which, to his consternation, was no more airworthy than his helicopters. Pressing on, he built his second aeroplane of 1910 and in this he achieved a 12-second flight. Sikorsky was airborne! From this point he went from strength to strength. His greatest pre-First World War achievement was the design of the mighty *Russky Vityaz*, (Russian Knight), an aeroplane best remembered as *Le Grand*. This was not merely the world's first aircraft with four engines, it was huge with a wing span of 91 feet 10 inches (28 metres). It made its maiden flight on 13 May 1913. On 2 August the *Russky Vityaz* managed a flight lasting 1 hour 54 minutes, while accommodating the pilot and eight passengers in a fully enclosed cabin. From this aeroplane Sikorsky developed the *Ilya Mourometz*, which first flew in January 1914 and carried no fewer than 16 passengers and a dog on a flight in February. Whether or not to prove its stability or for some other undefined purpose, the *Ilya Mourometz* was photographed with passengers promenading along the top decking of the fuselage while the aircraft was in full flight, a risky venture few would fancy today. At the outbreak of the First World War, the *Ilya Mourometz* was accepted by the Russian Army as a service machine and production got underway to equip the *Flotilla of Flying Ships* (EVK) with these aeroplanes for use as reconnaissance bombers. In January 1915 Sikorsky S-16 armed two-seaters were evaluated as escorts for the bomber.

Great success attended these bombers during the fighting but in 1917, with the Russian Revolution, Sikorsky left Russia for good to set up home and business in the U.S.A. Leaving behind virtually all his accumulated wealth, he found it difficult to get established in the United States. By 1924, however, he had produced a large transport plane, with the assistance of other Russians, under the company name Sikorsky Aero Engineering Corporation, founded on 5 March 1923. But the company's grand title hid a serious lack of finances. In October 1928 the company became a division of United Technologies and from then Sikorsky Aircraft produced a series of large flying-boats which served with Pan American Airways and others; the largest of these flying-boats helped establish several long over-water commercial routes. Yet, despite the excellence of the aeroplanes coming off the Sikorsky production lines, in the early 1940s Igor Sikorsky headed the company in a completely new direction. By 1944 Sikorsky had designed the world's first helicopter to enter full series production and service, but more of this later.

The first Soviet helicopter to fly with collective and cyclic pitch control was the TsAGI 1-EA, the work of Boris Nikolayevich Yuryev, which made its maiden tethered and manned flight in the summer of 1930. Powered by two 120-hp Le Rhône 9 engines (built under licence in the Soviet Union as M-2s), it had a single main rotor and no fewer than four anti-torque rotors at the front and rear of the fuselage. Two years later it was managing altitudes of up to 1,985 feet (605 metres) but is thought to have gone out of control during a test in mid-1933 and was severely damaged. Nevertheless, when rebuilt it continued to fly as an experimental machine from which data on rotor design could be gleaned.

It is entirely possible the 1-EA ended its flying career soon after the appearance of the similar 3-EA and more advanced 5-EA in 1933. The 3-EA was a trainer in design and, whether intentionally or unintentionally, it only ever flew in tethered form. The 5-EA adopted a six-blade rotor, probably for the first time on a flyable helicopter. This comprised three long blades and alternate short blades, the long blades only being capable of pitch control. Again using two M-2 engines, it was intended for experimental work only and reports published outside the U.S.S.R. gave it a range of about 13 miles (21 km) and, therefore, an endurance of about one hour. However, these figures almost certainly err greatly on the high side.

In 1936 the 11-EA appeared. This was basically of Cierva/TsAGI tandem two-seat autogyro configuration, with an aeroplane's fuselage and tail unit married to a new wing and six-blade rotor of 5-EA type but enlarged. Power was provided by an American Wright-Curtiss GV-1570F Conqueror Vee engine of 655 hp, mounted in the forward fuselage. This powered both the rotor and the two anti-torque propellers carried on the wing, the latter doubling as the source of thrust when the 11-EA (if required) was flown as an autogyro. Despite its closer association to the TsAGI's autogyros than the previous helicopters, the 11-EA was not very successful. Ivan Pavlovich Bratukhin, who had designed the rotors for this helicopter and the 5-EA, took over modifications of the 11-EA. In 1940 it flew in improved 11-EAPV form, with new blades and with the wing and propellers replaced by outriggers with anti-torque rotors. In this revised form it flew very well. It is known that the 11-EAPV was abandoned in the following year when the American engine began to show signs of wear. However, Bratukhin had already begun work on a twin-rotor helicopter, the official authorization for which was most certainly aided by Germany's *Operation Barbarossa*, the full-scale invasion of the Soviet Union, which had begun five days before.

Aeroplanes or helicopters?

While the events described above were taking place in the Soviet Union, the helicopter was beginning an important revival elsewhere. In 1926 an Italian by the name of Vittorio Isacco designed a very strange helicopter which he called the *Helicogyre*. The main feature of this machine was the use of an engine carried at the tip of each rotor blade to provide rotation without torque effect. It is likely that the only examples of Isacco's helicopter actually to be tested were those constructed under his guidance in Britain by Saunders-Roe in 1929 and in the Soviet Union during 1930. In the event his many years of work in various countries came to nothing, as his helicopter failed to lift.

Another Italian achieved greater success than Isacco, when in October 1930 the d'Ascanio heli-

Italian Vittorio Isacco began helicopter research in 1917, conceiving the idea of the Helicogyre. *The first machine built to his design was that constructed in Britain by Saunders-Roe in 1929, with an engine at the tip of each of four rotor blades. This was paid for by the British Air Ministry.*

copter was credited by the FAI of flying a distance of 0.67 mile (1.078 km) and in so doing achieving an endurance record of 8 minutes 45.2 seconds. It also set an altitude record of 59 feet (18 metres). These achievements, though seemingly low, were performed by the manned and untethered helicopter under the strict conditions laid down by the Fédération Aéronautique Internationale for world records.

This first d'Ascanio helicopter had been built in 1929 and was of the typical co-axial contra-rotating type, although an unusual feature was that each of the four wide-chord but tapering blades carried a trailing-edge stabilizing surface. These rotors and the three variable-pitch propellers (two mounted horizontally for lateral and longitudinal control and one vertically for anti-torque) were driven by a single 90-hp Fiat A.50 engine. This helicopter was subsequently acquired by the Italian Air Ministry. In 1934 d'Ascanio designed a follow-on machine with four-blade rotors but this was probably not completed. Just before the outbreak of the Second World War d'Ascanio's third machine was taking shape, featuring the modern single main rotor/anti-torque tail rotor layout. Whether this was ever flown is uncertain but after the war d'Ascanio worked with Piaggio & C. Societa per Azioni on a number of experimental helicopters at the Pontedora works.

Meanwhile, in October 1933, Belgium's Nicolas Florine took the world endurance record to nearly 10 minutes. By the beginning of 1935 the greatest FAI-accredited altitude reached by a helicopter was still below 200 feet (60 metres). It was again to France that historians turn for the next major development in helicopters, when on 26 June 1935 a co-axial twin-rotor helicopter flew which rapidly paled all earlier machines and their records into insignificance. Like Sikorsky (previously mentioned), Louis Breguet had not achieved much during his experiments with rotorcraft in the early years of this century and so had turned to aeroplanes. His company, formed before the outbreak of the First World War, produced the classic Breguet 14 and 19 biplanes and thereafter many varied types of aeroplanes from airliners to huge flying-boats. However, although prospering with aeroplanes, in the late 1920s his attention was drawn once again to rotorcraft. In 1931, having applied for patents covering means of helicopter stabilization, he formed a syndicate with René Dorand to develop a new helicopter. The result was the Breguet-Dorand *Gyroplane Laboratoire*, a single-seat experimental machine with a 350-hp Hispano-Suiza 9Q engine driving co-axial two-blade contra-rotating rotors. An aeroplane-type tail unit was mounted at the rear of the open fuselage structure, while the complete machine rested on single main wheels carried at the ends of wide outriggers, and nose and tail wheels.

The *Gyroplane Laboratoire* was at the crossroads of helicopter design. At first it was afflicted with some problems related to stability and control but it proved itself to be the first helicopter ever to perform really well in free flight. Three days before Christmas 1935 it managed a speed a fraction over 60 mph (98 km/h). Towards the end of the following year it attained an altitude of 518 feet (158 metres) and managed to remain airborne for over one hour while flying more

Opposite: After nearly three decades of fixed-wing aeroplane design Louis Breguet returned to helicopters and, with René Dorand, was responsible for the Gyroplane Laboratoire, *the first helicopter to fly well.*

Below: Original d'Ascanio helicopter of 1929 with two-blade rotors.

than 27 miles (44 km). It remained the subject of testing until fresh hostilities began in Europe in 1939 and was regrettably blown-up at Villacoublay by bombs dropped during an air raid from Allied aircraft in 1943.

It is curious that, given the excellent demonstrations, the *Gyroplane Laboratoire* was not the subject of greater contemporary acknowledgement. Indeed, this helicopter appears to have entirely escaped the notice of some of the most important aviation publications. However, this was not the case for the next milestone in helicopter development, which is credited to Germany.

Focke-Wulf Flugzeugbau AG, with works at Bremen, Flughafen and Johannisthal, was, by the mid-1930s, openly constructing both well-armed and multi-engined warplanes, for possible adoption by the reformed Luftwaffe, and trainers. Additionally, in 1931, Focke-Wulf had received a licence to construct Cierva Autogiros. Later Professor Heinrich Karl Johann Focke began work on a helicopter, no doubt having gained considerable experience of rotorcraft by way of the Autogiros. In 1934 he completed a model which flew well and thereafter set about organizing the fabrication of a full-size machine.

The result was the Fw 61, which took to the air for the first time one year to the day after the *Gyroplane Laboratoire* in France. As the basic structure it used the fuselage and vertical tail of one of the company's Fw 44 Stieglitz biplane trainers (with the forward cockpit faired over). The 160-hp Bramo Sh 14A radial engine in the nose powered the rotors and a tractor cooling-fan. On each side of the fuselage was attached an inclined pyramidal outrigger constructed of chrome-molybdenum steel tubing, which could be covered with balsa and Elektron material but were generally not. From each outrigger tip was carried a three-blade rotor, each blade made up of a tubular spar with plywood and fabric covering. The most important feature of the opposite-turning rotors was that each blade was double-articulated, with tangential oscillations limited by elastic tension. In other words this helicopter featured all the blade movements detailed in the first chapter for an entirely successful helicopter. The undercarriage was of nosewheel type and a braced tailplane was carried on top of the fin.

The first of two Fw 61s was civil registered *D-EBVU* and on its maiden flight remained airborne for 45 seconds. On 10 May 1937, having already achieved much longer flights, it became the first manned helicopter in aviation history to demonstrate an engine-off landing using autorotation (ground roll of 6 feet/2 metres). The second Fw 61 appeared in 1937 as *D-EKRA*. During 25–26 June 1937 Ewald Rohlfs flew the helicopters on separate occasions to establish new FAI-approved records: speed in a closed circuit of 76.151 mph (122.553 km/h), height of 8,002 feet (2,439 metres), duration of 1 hour 20 minutes 49 seconds, and distance in a closed circuit of 50.08 miles (80.604 km).

In 1937 Fraulein Hanna Reitsch, Germany's famous woman test pilot, became the first woman to pilot a helicopter when she took the controls. In February 1938 she flew an Fw 61 from Stendal to Berlin, a distance of 68 miles (109 km) and then demonstrated the machine inside the Deutschland Halle, where the floor area of 25,000 square feet (2,323 square metres) made possible fully controlled forward, backward and sideways flights without any assistance from wind. Later the same year another pilot took the FAI-accredited world distance record in a straight line to 143.07 miles (230.248 km), while in early 1939 a height of 11,243 feet (3,427 metres) was reached. The Fw

61 demonstrated that it could turn through 360 degrees in just 2.5 seconds while hovering and could reach 20 mph (32 km/h) in backward flight. It was estimated that, with the fairings fitted to the outriggers of the helicopter, a forward speed of 88 mph (123 km/h) might have been possible.

Clearly the Fw 61 was the world's first entirely successful helicopter and as such has overshadowed the achievements of the *Gyroplane Laboratoire*. The Fw 61's major shortcoming was its heavy weight, which was considerably higher than that of the Stieglitz biplane loaded for touring as a two-seater – and yet the Fw 61 was only a single-seater! In effect this meant that the Fw 61 was incapable of carrying a worthwhile payload, leaving only survey, observation or similar roles open to it. However, by now the Fw 61 had been under the patronage of the Focke-Achgelis company for some years, a company founded by Heinrich Focke with Gerd Achgelis after leaving Focke-Wulf. It was, therefore, this company that continued the development of the Fw 61 and, using the experience, produced the first-ever helicopter to go into limited production, as the much larger Fa 223.

As the following chapter recounts, Allied bombing kept the Fa 223 out of major series production for wartime service with the Luftwaffe and only a handful were ever used. Such action altered what might otherwise have been the course of history, allowing the first helicopter to enter full production and military service to be an American type.

Meanwhile back in Britain in 1937, G. and J. Weir, a Scottish company that had constructed Cierva-type Autogiros, decided to devote itself entirely to helicopters. Its two-seat W.5 (60-hp Weir engine) was influenced by the Focke-Wulf Fw 61. The W.5 first flew on 7 June 1938 at Dalrymple, making it the first partially successful British helicopter. Flown by Raymond Pullin, the company's chief engineer, it had cyclic pitch control but no collective pitch control, vertical flight being controlled by increasing or decreasing the engine's rpm. As a result of these flight tests, the Air Ministry ordered the follow-on W.6.

The Russian emigrant Igor Sikorsky had, as previously described, established a new life for himself in the U.S.A. and founded a company which became part of United Technologies then United Aircraft Corporation. Sikorsky had never lost interest in the helicopter and had, in 1931, taken out a patent to cover a helicopter with major innovations. Perhaps the most important of these innovations was the selection of a single main rotor and a small vertical anti-torque rotor, though this idea was not entirely new: Achenbach, in 1874, conceived a design using an anti-torque rotor and much later, in the 1920s, the Dutchman, von Baumhauer also hit upon the concept of a single main rotor and an anti-torque secondary rotor.

As the Engineering Manager of UAC's Sikorsky Aircraft division, Igor Sikorsky approached the management in the latter part of 1938 with the recommendation that design and development should begin of a direct-lift machine. With his early helicopter experiments behind him and nearly three decades of experience in the design of aeroplanes, it could not have been too much of a surprise

Above: VS-300 with a tail rotor and small outrigger-carried horizontal rotors.

Opposite top: Hanna Reitsch in control of Focke-Wulf Fw 61 D-EKRA *in 1937.*

Opposite below: With Igor Sikorsky at the controls, the VS-300 in original configuration lifts from the ground during its first tethered flight on 14 September 1939.

when approval was given for the project.

In April 1939 Sikorsky Aircraft merged with Chance Vought to become Vought-Sikorsky, still a division of UAC. Thus it was that Vought-Sikorsky began a courtship with helicopters that would eventually lead to the separation of Vought and Sikorsky and the emergence of one of the world's greatest helicopter producers.

The Vought-Sikorsky VS-300 appeared in 1939 and was very different from any helicopter to be seen in Europe. The fuselage comprised an open structure of steel tubes forward and a boom aft carrying both a ventral fin and the tail rotor. On the forward structure sat the pilot, with a 75-hp Lycoming four-cylinder engine to his rear under the main rotor pylon. The 28-foot (8.53-metre) diameter three-blade main rotor had cyclic pitch control. Sikorsky himself piloted the VS-300 on its first tethered lift-off, on 14 September 1939, and other brief tethered lifts followed in November with weights suspended from the airframe to assist stability. The lack of adequate control over flight was put down to the cyclic control system and so the VS-300 was taken away for modification.

In its revised form the VS-300 used its main rotor for lift only. The rear boom gave way to a new welded structure of steel tubes, the extreme tail of which supported a small vertical anti-torque rotor. However, slightly further forward on the open fuselage structure were attached outriggers on which two small horizontally mounted rotors provided longitudinal control (pitch in same direction) and lateral control (pitch in opposite direction). Flying resumed and on 13 May 1940 it made its first untethered free flight to become the most successful helicopter outside Germany. In the following year the Lycoming gave way to a 90-hp Franklin engine. In April the VS-300 donned two large rubber flotation bags to allow a trial ascent from water. Though not the first waterborne helicopter (Papin and Rouilly of 1915), in this form it became the first helicopter to achieve a successful take-off from water. On 6 May Igor Sikorsky flew the helicopter to a new world endurance record of a little over 1 hour 32 minutes, thereby eclipsing the German Fw 61's record.

From June 1941 the VS-300 underwent important modifications: the outriggers were removed, cyclic pitch for lateral control was adopted and a horizontal tail rotor was used for longitudinal control. In December of that year a new 30-foot (9.14-metre) diameter rotor with full cyclic pitch control was fitted, the horizontal tail rotor removed and a vertical anti-torque rotor substituted. Thus the classic helicopter configuration had come into existence. Further changes to the VS-300 involved the substitution of a 150-hp Franklin engine and the use of fabric to cover part of the airframe. The VS-300 continued to fly in 1942 and in the following year was retired to the Edison Museum. Its maximum speed and range were probably 50 mph (80 km/h) and 75 miles (120 km) respectively.

A helicopter industry was now about to be born in the U.S.A. At the same time, German attempts to establish an industry were virtually bombed away by the Allies, and the Soviets were struggling with several designs, but were unable to make significant progress until peace returned.

American Gains - German Losses

Although France had played a major role in the early development of the helicopter, once the Second World War began and France became first a nation under siege and then occupation, helicopter advancement was forced backstage. Whilst French engineers managed to make some inroads into advanced technologies under the noses of the occupying Germans (an obvious example being that of gas turbine engines), further development of helicopters suffered. By contrast, both Germany and the U.S.A. produced helicopters suitable for full production and service during the war, although German efforts to get large numbers into military service failed almost entirely as the result of Allied air attacks.

The two greatest German exponents of helicopters during this period were Flettner and Focke-Achgelis, both concerns developing machines that received large production orders. Anton Flettner preceded Heinrich Focke as a designer of rotorcraft, his earliest machine having appeared in 1932. This curious helicopter worked on the same lines as the contemporary Italian Isacco, with a small Anzani engine mounted on each rotor blade to provide drive. The only advantage of this layout was the absence of torque effect but the machine was lost before any free flight could be attempted. The fact that it flew at all, albeit tethered, rates it as more successful than the Isacco.

Having lost this helicopter Flettner turned to the autogyro, producing the very advanced-looking Fl 184 with a fully enclosed cabin for the pilot. This machine, the prototype of which was destroyed by fire in 1935 before it could be evaluated, was followed by a fairly unsuccessful helicopter-cum-autogyro known as the Fl 185, which had a single main rotor and pusher anti-torque/propulsion propellers carried on booms from the fuselage sides.

The Fl 184 and Fl 185 used nose-mounted Sh 14A radial engines: in the case of the Fl 184 it was intended to provide forward propulsion via a conventional two-blade propeller: in the Fl 185 it drove the rotor and pusher propellers plus a nose fan as on the Fw 61. A similar nose-mounted engine was selected for Flettner's Fl 265 helicopter, of which six prototypes had been ordered in 1938 for evaluation by the German Navy. As with the Fl 184, the pilot was accommodated in a small cabin and an aeroplane-type tail unit was adopted. The main difference with the Fl 265 was its rotor system which comprised two opposite-turning two-blade intermeshing rotors mounted with their axes close together and synchronized to avoid collision. The curious fact that German helicopters virtually always adopted a twin-rotor layout, even when of small proportions, is hard to explain.

Critics of Flettner's layout considered intermeshing blades to be less efficient than the more traditional twin-rotor configuration using outriggers to carry non-overlapping blades, but Anton Flettner decided that the reduction of drag through abandoning outriggers offset any other considerations. The first Fl 265 prototype appeared in May 1939 and others quickly followed for evaluation as aircraft to be used from ships of the German Navy and as Army co-operation aircraft. Of these only two were lost, one in a worrying accident when the rotors struck each other and the other after running out of fuel during a trial. In most aspects of flight the Fl 265 proved successful and it was viewed by the German Navy as a good prospect for anti-submarine and reconnaissance duties with the fleet.

Although intended for series production, the Fl 265's career was abruptly curtailed by the development of the even better Fl 282 Kolibri, which had the advantage of being a two-seater with the Sh 14A engine carried behind the seats to allow a much improved forward view. No fewer than 45 prototype and pre-production helicopters of this type were ordered; the rapid development of the helicopter was of great importance since France's capitulation left the way open for Germany's more ambitious plans to invade Britain and the Soviet Union.

Early prototypes appearing from 1941 were completed as single-seaters for trials, with both open and enclosed accommodation. In 1942 the cruiser *Köln* was used for trials at sea, with the helicopter operating from a gun-turret platform. But production of the Fl 282 was proving painfully slow and by 1943, when prototypes were cleared for experimental shipborne operations to protect convoys in the Mediterranean and Aegean seas, less than half of the helicopters originally ordered in 1940 had been delivered. In 1944 BMW received the go-ahead for the production of 1,000 machines but not a single

Flettner Fl 262 Kolibri prototype V19, not one of the three captured by the Allies in flying condition but nevertheless of particular interest as it was completed with the nose cowling (rotors missing here).

example left its factories, the direct result of Allied air raids. By the close of the war only 24 Fl 282s had been delivered, three of which had joined the historically famous Luft-Transportstaffel 40 unit. The maximum speed of the Fl 282 was 93 mph (150 km/h).

At the time of the German capitulation, Flettner had been developing a new helicopter as the Fl 339, which was intended to carry 20 troops. But even this helicopter weighed less than the wartime Focke-Achgelis Fa 223 Drache that had managed limited service with Luft-Transportstaffel 40.

Following the success of the Fw 61, the German airline Deutsche Luft-Hansa had no reservations about ordering a six-seat development of generally similar but larger configuration for commercial use. The prototype of this commercial helicopter appeared in 1939 as the Fa 266 Hornisse. This was the first helicopter in history to have been designed and built for transport duties and the first to be completed to an order placed by a commercial airline. However, the war interfered with Luft-Hansa's plans and, before the Hornisse had a chance to fly, it was modified to suit military roles.

As the Fa 223 Drache, it began tethered trials in the early part of 1940 and achieved its initial free flight in August that year. As with the Flettner, it was less stable at low speeds and had the added vice of rotor vibration at high speed, but was able to demonstrate remarkable lifting ability. Towards the end of October 1940 the Fa 223 attained an altitude of 23,300 feet (7,100 metres), far exceeding any previous height by a helicopter.

Thirty pre-production Fa 223s were ordered for evaluation and experimental operation. Intended roles included anti-submarine duties (armed with the standard MG 15 nose gun and two 250-kg/550-lb bombs or depth charges), transportation (using the cavernous main cabin or with a slung load of normally 1,760 lb/800 kg) and rescue. The 1,000-hp Bramo 323 Q3 Fafnir engine drove two 39-foot 5-inch (12-metre) diameter three-blade rotors carried on huge outriggers. The Fa 223's gross weight was considerably higher than that of the Fa 266, largely because of the more powerful engine (the Fa 266 was to use an 840-hp engine), and its maximum possible speed was marginally below that of the commercial helicopter's intended 118 mph (190km/h). However, the Drache was normally not flown above 75 mph (120km/h).

Actually less important to Germany's war effort than the Fl 282 because of the very small number delivered for operational service, the Fa 223 nevertheless had the potential of being the most useful helicopter of any nation during the 1939–45 war. The second Fa 223 prototype flew in July 1942. Acceptance of the helicopter for service resulted in an order for 100 full-production Fa 223Es but the Allies saw to it that this could not be fulfilled. The first Fa 223 production line at the Bremen factory was destroyed by Allied bombers before any further machines could be delivered; production moved to Laupheim. Two years after the second prototype had flown, only a further eight Fa 223s had been completed and tested, of which all but two were destroyed that month in another Allied air attack. As the Allied forces advanced following the D-day landings, a third and final production line was set up in Berlin which had the capacity to produce hundreds of helicopters a month when operating at

Captured Focke-Achgelis Fa 223 Drache in USAAF markings, photographed some weeks before the German surrender.

maximum output. Only a single Fa 223 was completed at Berlin before Germany's capitulation. The Fa 223E was intended primarily for Army support as the German situation in Europe deteriorated but it is believed that just three were operated by Luft-Transportstaffel 40.

Of two Fa 223s captured in flyable condition by the Allies, one was flown to Brockenhurst, Hampshire, England, in September 1945 by its German crew of three, making the first-ever helicopter crossing of the English Channel. This must have been one of the two Laupheim helicopters that had survived the July 1944 air attack, but its luck ran out in October when it was destroyed in a flying accident.

In Czechoslovakia, the former Avia factory at Cakovice (enlarged by the Germans when they occupied the area) proved ideal for the manufacture of German aircraft to arm the postwar Czech air force. It was here that two Fa 223s were completed from abandoned components for trials as VR-1s, along with a number of fixed-wing aircraft including Messerschmitt Me 262s and Me 109Gs. Another Fa 223 was assembled in France after the war by SNCASE, under the supervision of Professor Heinrich Focke. First flown on 23 October 1948, the French version was virtually identical to other Fa 223s and was known as the S.E.3000. However, it was converted during manufacture into a six-seat commercial helicopter, with a four-passenger cabin aft of the flight deck. The S.E. 3000 had a gross weight of 9,445 lb (4,284 kg), slightly under the maximum weight of the German service machine, and could fly at 113 mph (182 km/h). It had a maximum rate of climb (not vertical) of 1,575 feet (480 metres) per minute and a range of 230 miles (370 km).

In Chapter Two the Focke-Achgelis rotor-kite was detailed. It

French-assembled Fa 223 known as the S.E.3000, the work supervised by Professor Focke. It featured a four-passenger cabin to the rear of the flight cabin.

is interesting to note that the company also developed a rotor-kite version of the DFS 230 transport glider, although this (the Fa 225) remained a prototype. Of various projects begun and abandoned by Focke-Achgelis, one which deserves mention is the Fa 284, which would have been a twin-rotor flying crane. However, other strange helicopters from different sources were developed to actual prototype stage for possible service with the German forces and these are worthy of notice. Austrians Bruno Nagler and Paul Baumgartl both produced ultralight pilot-only helicopters that ran on tiny engines, and AEG flew a co-axial twin-rotor helicopter powered by a 200-hp electric motor that was intended for tethered observation flights.

The concept of tip-jet helicopters was also re-invented in Germany during the war. The resulting Doblhoff/WHF 342 prototypes have often been spuriously pronounced as the world's first jet-driven helicopters. However, the Doblhoff/WHF 342 V1 first prototype, when it made its maiden flight in early 1943, *was* the first jet-powered helicopter to fly successfully. It was powered initially by a 62-hp Walter Mikron II four-cylinder engine, which drove an air compressor which in turn passed air through the hollow rotor blades to tip-jets where, fuel having already been added, the mixture was burned to produce thrust. Not only was the rotor system simple, it produced no torque effect. Development of the helicopter by Baron Friedrich L. von Doblhoff and his assistants progressed through the single-seat V2 and V3 prototypes (the V2 was the V1, which had been damaged during an air-raid and rebuilt with a 90-hp Mikron) and led to the most important prototype, the two-seat V4. Powered (like the V3, which destroyed itself by excessive vibration) by a Sh 14A engine, the V4 used this to drive the air compressor for vertical, hovering and slow forward flight and a pusher propeller carried at the rear of the fuselage nacelle for cruising flight, at which time the rotor autorotated. Why Doblhoff had turned his helicopter into a helicopter-cum-autogyro is simple to explain. He had hoped that his small machine, with its promise of simplicity, would fulfil the German Navy's requirements for a ship- and submarine-borne observation helicopter. But flying trials with the early examples had shown them to be very thirsty for fuel. In the event V4 was still in its early test phase when Doblhoff and his assistants fled the advancing Soviet forces, taking with them the V2 and V4. These were captured by U.S. forces. It is not known what speed V4 was capable of achieving, as Doblhoff had not attained a figure. However, it would almost certainly have been equal to other contemporary German two-seat helicopters. Friedrich von Doblhoff joined the American company McDonnell after the war and helped in the

Wiener Neustadter Flugzeugwerke-built Doblhoff 342 V4.

development of the XV-1 experimental convertiplane of 1955 that adopted rotor-tip pressure-jets. Of his assistants, Stepan became an employee of the British Fairey company and his work aided the development of the Rotodyne.

The United States of America takes the initiative

Although the Fa 223 Drache might have become the most important helicopter of the Second World War, this accolade went instead to the U.S.'s Sikorsky R-4. Having arrived at a satisfactory configuration for the VS-300 by 1941, although further changes were made to this helicopter thereafter, Vought-Sikorsky received a contract to supply the U.S.A.A.F. with a single example of a two-seat derivative for evaluation as the XR-4. The chosen power plant was a 165-hp Warner R-500-3, carried inside the flat-sided fuselage that was constructed of welded steel tubes with metal covering forward and fabric aft. The engine powered both the 36-foot (10.97-metre) diameter main three-blade rotor and the tail rotor and, in essence, the XR-4 was little more than a scaled-up VS-300 of less rounded shape. The two crew were accommodated side-by-side in a well-glazed enclosed cabin at the nose of the aircraft, affording excellent forward, downward and sideways vision.

The XR-4 made its maiden flight on 14 January 1942 and in May was flown in stages from the Stratford, Connecticut, factory to Wright Field, Dayton, Ohio, for evaluation. This delivery flight represented something of a milestone in itself, covering a distance of 760 miles

A Sikorsky YR-4B takes off from a 10th Air Force airfield in Burma.

(1,220 km) and flown in various weather conditions without a hint of mechanical trouble. Later that year an additional 30 helicopters were ordered for U.S.A.A.F. evaluation, each with a more powerful 180-hp Warner R-550-1 seven-cylinder radial engine to drive a 38-foot (11.58-metre) diameter three-blade rotor.

The first three development aircraft were designated YR-4As and were basically similar to the XR-4, the latter receiving the more powerful engine in 1943 to become the XR-4C. The remaining 27 development aircraft were YR-4Bs, differing mainly in having larger cabins. Meanwhile, in January 1943, Sikorsky separated from Chance Vought to become the Sikorsky Aircraft division of UAC, a move considered appropriate to allow Chance Vought to continue the production of warplanes and Sikorsky the manufacture of helicopters. Sikorsky Aircraft moved from Stratford to Bridgeport to take up a newly leased factory. The flying ground at the factory was the

The first released photograph of Sikorsky R-4B full-production helicopters during construction at Bridgeport. The rear fuselage structures moved down the left side of the factory to be married to the front sections, then to have the engines installed before moving up the final assembly line on the right.

An RAF Hoverfly I on display at Abingdon (KK995), one of 45 supplied under Lend-Lease.

smallest for aircraft in the world but was nevertheless adequate for helicopters.

Of the development aircraft, three went to the U.S. Navy in 1943 as trainers under the designation HNS-1 and seven were transferred to the R.A.F. as Hoverfly Is. HNS-1s and Hoverfly Is were the first-ever helicopters to enter service with these forces. Six of the R.A.F. aircraft joined the Helicopter Training Flight based at Andover in early 1945 and one went to No. 529 Squadron. U.S.A.A.F. YR-4s were used for both training and service evaluation, including trial flights in the tropical conditions of Burma and the cold conditions of Alaska. Also, in May 1943 the U.S.A.A.F. assisted the War Shipping Administration and Coast Guard in demonstrating the operation of helicopters from ship platforms. The ship used was the tanker *Bunker Hill*, which had a space cleared on its cargo deck only 14 feet (4.25 metres) greater than the 38-foot (11.58-metre) diameter of the main rotor. Twenty-four lift-offs and landings were performed with the ship anchored and moving at Long Island Sound. Because of the success of these tests the U.S. Navy decided to acquire a further 20 helicopters from the U.S.A.A.F., giving the responsibility for further trials and pilot training to the U.S. Coast Guard, which also intended to use the helicopter for air-sea rescue operations. Unfortunately it was decided that the helicopter was not sufficiently developed for an anti-submarine role.

The Navy's main allotment, plus 45 helicopters supplied to the R.A.F. under Lend-Lease arrangements, were from a batch of 100 full-production Sikorsky R-4Bs. As these were in production in 1944, the R-4B was the world's first helicopter to enter full production and then service. The R-4B was powered by a 200-hp Warner R-550-3 engine and could fly at 75 mph (121 km/h), a considerably lower speed than attained by the lighter prototype. It could climb to an altitude of 8,000 feet (2,440 metres) in 45 minutes, which repre-

Top: USAAF Sikorsky R-4B.

Above: Second prototype Sikorsky XR-5, a major improvement over the earlier R-4.

sented its service ceiling. Normal range was 220 miles (354 km). Either a three-wheel undercarriage could be fitted, using widely spaced shock-absorbing struts supported on steel pyramids, or twin low-pressure rubberized pontoons which were useful for water operations. Optional service equipment included an externally carried stretcher. In 1948 those R-4Bs still in service with the U.S.A.F. were redesignated H-4Bs but these were quickly superseded by the wartime-developed Sikorsky S-51. R.A.F. R-4Bs were also retired soon after the war, although a small number continued flying as radar calibration aircraft and some others performed experimental tasks well into the 1950s.

Development of the R-4 type helicopter by Sikorsky was very rapid indeed and before the end of the Pacific War two other helicopters appeared that offered higher performances and varying accommodation. The first of these to fly was the R-5, the development of which had begun soon after the R-4 had proved itself. The R-5 became the most important early production Sikorsky helicopter. Five prototypes were ordered for evaluation against a U.S.A.A.F. requirement for an observation helicopter; the first XR-5 took off for the first time on 18 August 1943.

In configuration the R-5 was totally new, although it retained the single main rotor and tail rotor layout that became accepted for the great majority of helicopters and remains so to this day. A 450-hp Pratt & Whitney R-985-AN-5 Wasp Junior radial engine was mounted under a new 48-foot (14.63-metre) diameter three-blade main rotor; the increased power was necessary to lift the much larger fuselage. Indeed, the R-5 had a gross weight nearly twice that of the R-4. The fuselage itself was basically composed of three sections. The nose section accommodated the crew of two in tandem, with the pilot to the rear in an aluminium structure that was heavily glazed with Plexiglas windows. The centre section was a welded steel tube structure with plastic-impregnated moulded plywood covering, and the tailboom was a wooden monocoque.

The XR-5s were followed by 26 YR-5A development aircraft and 34 full-production R-5As for the U.S.A.A.F. The latter had provision for several items of operational equipment, including cameras and up to four stretchers carried in pairs on the fuselage sides. R-5As became the first helicopters used by the Air Rescue Service for search and rescue, normally with provision for two stretchers (the mounts projecting from slits in the fuselage sides). Twenty-one R-5Ds were produced by modifying R-5As to use 600-hp Pratt & Whitney R-1340 Wasp engines. Each R-5D also included provision for a rescue hoist, extra fuel for extended range and the adoption of a nosewheel to the standard three-wheel gear. In addition, five of the development aircraft were redesignated YR-5Es after receiving R-5A type dual controls.

In 1947 the U.S.A.F. acquired 11 four-seat R-5Fs but in the following year the helicopter's designation was changed to H-5. Therefore, the last examples for the U.S.A.F., all built by 1951, were delivered as H-5Gs (39) and amphibious H-5Hs (16). The following chapter outlines some of the work carried out by R-5s during the Korean War and it is sufficient to say here that the last R-5s in U.S.A.F. service were still flying in the 1960s. The next

Sikorsky R-5 from the Elmendorf Field detachment of the 10th Rescue Squadron, Air Rescue Service, USAF, photographed during a simulated Arctic rescue mission.

chapter also describes the British variant, built by Westland, and commercial S-51s. In addition, the U.S. Navy received about 90 R-5 helicopters under the HO3S designation. The maximum speed of the R-5A was 90 mph (145 km/h), which rose with subsequent variants to 106 mph (171 km/h).

The other wartime Sikorsky helicopter was the R-6, which was little more than a refined R-4 with a new, more modern and longer fuselage accommodating two crew side-by-side with dual controls for observation duties. The helicopter's rotor and transmission were pure R-4, although power was provided by a 225-hp Lycoming O-435-7 engine in the XR-6 prototype (that first flew on 15 October 1943) and by 240-hp Franklin O-405-9 six-cylinder engines in the 26 development YR-6As and 193 full-production R-6As for the U.S.A.A.F. The Nash-Kelvinator Corporation was made responsible for R-6 production, as Sikorsky concentrated on the more demanding R-5. Of these, the U.S. Navy acquired three development aircraft and 36 full-production R-6As from the U.S.A.A.F, under HOS designations. Naturally, these were identical to Army models. A further 26 found their way to Britain after the war, to become Hoverfly IIs with the R.A.F. and F.A.A. However, the R-6A was a less important helicopter than the contemporary R-5. Optional equipment included two externally carried stretchers. The R-6A had a maximum speed of 100 mph (161 km/h), weighed almost the same as the slower R-4 and could remain airborne for up to 5 hours.

Below: Sikorsky S-51 in civil markings.

Other American developments

Partly due to the war, the helicopter became the focal point for considerable development in the U.S.A. in the early 1940s with some astonishing results. A name that

Right: The first Sikorsky XR-6A prototype, one of five that preceded the YR-6As and R-6As. XR-6As used the same type of engine as later production models.

ranks with that of Sikorsky is Bell. The Bell Aircraft Corporation was founded in 1935 by Lawrence D. Bell, R. P. Whitman and Robert J. Woods, all of whom were ex-Consolidated Aircraft Corporation employees. This company produced warplanes of unusual design and was also responsible for the first-ever U.S. jet fighter. However, some months before the United States entered the Second World War, Bell began work on a helicopter as a private venture and by the middle of 1943 the first Model 30 two-seater prototype was flying. This spawned the postwar Bell Model 47, the first major helicopter from Bell and one which became the subject of mass production in several countries. Model 47 production finally came to a close with the delivery of the last Italian-produced examples in 1976. Many historically important events are associated with the Bell Model 47 and these are described in the following chapter.

In 1944 the first-ever Hiller helicopter appeared as the experimental XH-44, a cabin single-seater with a 125-hp Lycoming engine driving co-axial contra-rotating two-blade rotors. Demonstrated in public in August that year, it was probably the first successful co-axial helicopter. On 7 August 1944 Kellett, the former autogyro manufacturer, flew the prototype of a two-seat helicopter with intermeshing contra-rotating blades, the first such machine to be

The first tethered flight by the first Bell Model 30.

Right: The second Bell Model 30 was completed with a car-type cabin.

Below: The first Model 30 was flown without the fuselage structure covered, partially covered and with faired undercarriage struts, and with the fuselage fully skinned. Here the Model 30 demonstrates its capabilities as an agricultural aircraft.

flown in the U.S.A. Tested as the XH-8 by the U.S.A.A.F. (two built), its 245-hp Franklin engine allowed a performance similar to that of contemporary Sikorsky helicopters.

On 2 November 1944 the Landgraf Helicopter Company of California flew an experimental cabin single-seater as the Model H-2. This incorporated several unusual features such as a cyclically controlled aileron system on the tips of the rotor blades and overlapping rotor-blades rigidly attached to hubs and carried on fully faired canted booms.

From the P-V Engineering Forum, founded by a group of engineers mutually interested in rotorcraft, came the single-seat PV-2, powered by a 90-hp Franklin engine. This helicopter is particularly remembered for two reasons. It was only the second helicopter to be demonstrated in public in the U.S.A., in September 1943 at Washington National Airport, and it brought the name of Frank N. Piasecki to the forefront. The PV-2 actually made its maiden flight on 11 April 1943. Piasecki was President of the P-V Engineering Forum until the name of the organization changed in 1946 to Piasecki Helicopter Corporation. On 1 February 1944 P-V Engineering received a contract from the U.S. Navy for a prototype of the PV-3 'Dogship', a tandem rotor helicopter capable of accommodating the crew and 10 passengers. As the Navy's XHRP-X, this prototype flew in March 1945, followed by two XHRP-1 prototypes (one for static and dynamic tests) and then in 1947 by the first production examples of Piasecki's famous 'Flying Banana'. Powered by a single 600-hp Pratt & Whitney R-1340-AN-1 Wasp engine, driving two three-blade rotors, the HRP-1 was the first fully successful twin-rotor helicopter to

First public demonstration of the Hiller XH-44 at San Francisco in August 1944.

enter full production and service. More details of this helicopter and its U.S.A.F. derivatives follow in the next chapter.

Of lesser known helicopters of American wartime origin can be included the Platt-Le Page XR-1 and XR-1A, the first helicopters to receive a U.S.A.A.F. 'R' for rotary-wing designation. Two were ordered under U.S. government contract for evaluation and were very similar indeed in configuration to the German Focke-Achgelis Fa 223. Each had a single 450-hp Pratt & Whitney R-985-AN-1 Wasp Junior engine driving 30-foot 6-inch (9.3-metre) diameter three-blade rotors carried on fuselage outriggers, and differed only in the amount of glazing around the tandem two-seat nose cockpit. These aircraft completed successful trials with the U.S.A.A.F.'s Air Technical Service Command at Wright Field and were accepted by the U.S.A.A.F. However, no production followed.

Bratukhin in the Soviet Union

There can be little doubt that the design of the American Platt-Le Page was influenced to some degree by experiments in Germany. The same is true of the most important helicopter experiments of the wartime Soviet Union, conducted under the supervision of Ivan Pavlovich Bratukhin who had already assisted in the development of helicopters at the TsAGI.

In mid-1941, just five days after Germany launched *Operation Barbarossa*, the invasion of the Soviet Union, Soviet authorities gave Ivan Bratukhin the green light to develop a large twin-engined and twin-rotor helicopter that had several possible military uses including transport and artillery spotting. As built, the Omega had a metal tubular fuselage structure that was fabric covered. The nose had a glazed panel and the crew of two sat in tandem in a high-mounted cockpit. Each open-girder outrigger structure from the fuselage sides supported a 220-hp MV-6 radial engine driving a 23-foot (7-metre) diameter three-blade rotor via a long mast. A T-tail was adopted and the complete machine rested on four wheels.

Left: The P-V Engineering Forum PV-2, flown by Frank Piasecki. It was only the second helicopter to be demonstrated in public in the USA.

Right: The first US helicopter with intermeshing contra-rotating blades was the Kellett XH-8.

Below: The Piasecki XHRP-1, the first of the famous 'Flying Bananas', was derived from the P-V Engineering Forum XHRB-X 'Dogship', which had been built to the first ever US military contract for a helicopter.

Right: The Platt-Le Page XR-1 and XR-1A (illustrated) were the first helicopters to receive USAAF 'R' for rotary-wing military designations.

Above: The Bratukhin Omega II, first displayed in public at Tushino on 18 August 1946, its two massive nacelles housing MG 31F engines.

In view of the urgency of the situation, the Omega was ready for initial tethered trials in August. However, outside events soon caused a halt to these while a safer location was found. Further tethered lifts followed in 1942, leading to free flights in 1943. As testing revealed the original power plant to be less than suitable, the Omega subsequently underwent modification that included the installation of 330-hp Kossov MG 31F engines. Flying resumed in September 1944, still with K.I. Ponomaryev at the controls, but the helicopter was out of action for most of the first half of 1945 following an accident. In August 1946 the Omega II was shown in public for the first time during the Soviet Aviation Display at Tushino. Because of its heavier gross weight in modified form, the Omega II could only manage 93 mph (150 km/h) but was a major milestone in Soviet helicopter design. For his achievements, Bratukhin eventually received a Stalin Prize.

Meanwhile, by 1944 the Omega had flown sufficiently well to warrant the authorization of development helicopters for experimental service with operational units. The new G-3 was intended to be virtually identical to the earlier helicopter, except for the adoption of two American Pratt & Whitney R-985-AN-1 Wasp Junior engines (as, incidentally, fitted to the Platt-Le Page) which had been made available through Lend-Lease. Two prototypes appeared in 1945, followed by a handful of production machines, but these appeared only after hostilities had ended. Although the G-3 had an even heavier gross weight than the Omega II, the increase in power allowed a speed of 105 mph (170 km/h). These were not the last of the Bratukhin twin-rotor helicopters, as the next chapter reveals.

An Industry Grows

The helicopter, tested thoroughly during the Second World War but barely put to any practical military use, was readily accepted into postwar military service. Establishing the helicopter on the commercial scene proved somewhat slower.

Historically the first major events concerning helicopters of the postwar era were, ironically, on the commercial side. On 8 March 1946 the Bell Model 47, a small two-seat machine developed from the wartime Model 30, was awarded the first U.S. commercial licence (*NC-1H*) and the first-ever Type Approval Certificate for a commercial helicopter anywhere in the world. The earliest production Model 47s were powered by 178-hp Franklin engines and had car-type cabins for two persons. Examples delivered to the U.S. Navy as HTL-1s and U.S.A.F. as YR-13s (later H-13s) in 1947 were the first of a huge number to be acquired by the U.S.A.F., U.S. Navy, U.S.M.C., U.S. Army and U.S. Coast Guard (HUL designation) over many years, with production for the U.S. forces ending in 1968 and all manufacture ceasing in the U.S.A. in 1974. Most later versions adopted the more familiar 'goldfish bowl' canopy, had more powerful engines and accommodated three or four persons. The U.S. Navy found the Model 47 ideal for pilot training and general duties, while the U.S. Army models, known as H-13 Sioux, proved suitable for many duties including liaison, casualty evacuation and observation.

The initial civil version was the Model 47B, either with the car-type body or as the Model 47B-3 with an open cockpit. The 'goldfish bowl' canopy was introduced on the Model 47D. It is interesting to note, at this point, how Bell decided to demonstrate the helicopter's worth as a commercial machine. In early 1947 Larry Bell suggested that the company's chief test pilot, Tex Johnson, with Dick Stansbury, should take a Model 47B-3 and

The first type of helicopter to be awarded a US commercial licence and the first to receive a Type Approval Certificate for a commercial helicopter was the Bell Model 47. The very first helicopter to be so registered was NC-1H.

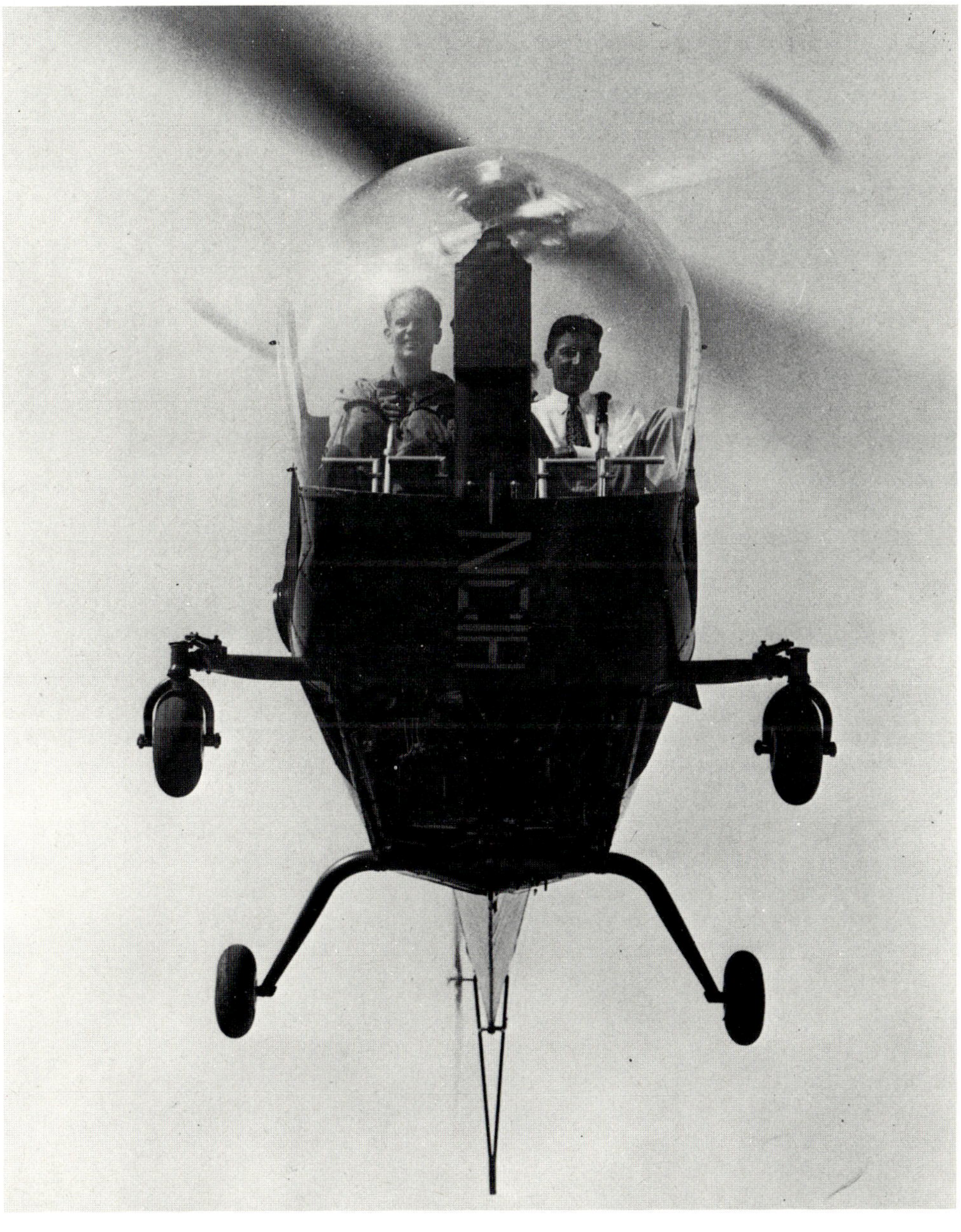

discover whether it could make money. The pair first undertook agricultural demonstrations in Arkansas and then, while en route to crop-dust in the Rio Grande Valley, took on the job of transporting geophysicists of the Shell Oil Company between locations in the Louisiana marshes, a job which required a second Bell helicopter. This was undoubtedly the first introduction of the helicopter to the oil industry and, incidentally, eventually led to the establishment of one of the world's most important helicopter-operating companies, that of Petroleum Helicopters Incorporated.

By 1948 Bell Model 47s were operating commercially in several countries, including Argentina where they helped combat the annual locust plague for the first time and thus saved crops worth millions of dollars from ruin. The Model 47 was the first helicopter

Left: Bell Model 47D-1s were used by Sabena in 1950 to inaugurate the first helicopter air mail service in continental Europe. This 47D-1 lands at Brussels.

Below left: Westland-built Sioux (AH Mk 1 in British Army service). Based on the Model 47G-3B-4, these were constructed under licence from the Italian Agusta company that held the European rights from Bell.

Below: Bell HSL-1, the company's only twin-rotor helicopter, seen here in US Navy service towing a vessel.

licensed in the U.S.A for crop-dusting, the helicopter's slow flying speed and rotor downwash helping to establish rotorcraft in agriculture. The authorities were also quick to understand the special nature of helicopters and the CAA revised its flying regulations to permit rotorcraft to fly below an altitude of 1,000 feet (305 metres) over populated areas. Production of the Model 47 was also undertaken by Westland in the U.K., Kawasaki in Japan and Agusta in Italy. Other early Bell helicopters included the Model 48 tactical helicopter, of which 10 service evaluation YR-12Bs were purchased (after prototypes) by the U.S.A.F. in the 1940s, and the Model 61, Bell's only tandem-rotor helicopter, of which 50 production examples were delivered to the U.S. Navy during the 1950s as specially designed anti-submarine helicopters. Each Navy Model 61, designated HSL-1, was powered by a 2,400-hp Pratt & Whitney R-2800-50 engine and was expected to carry early air-to-surface missiles.

Many of the first mail services by helicopter were conducted by the civil version of Sikorsky's R-5/H-5, known as the S-51. This was the first Sikorsky helicopter to be licensed by the CAA for commercial operation and, like the Model 47 from Bell, went into production in 1946. Related to the U.S.A.F's H-5F, it was a four-seat helicopter with the pilot seated centrally in the forward part of the enlarged cabin and the three passengers occupying a bench seat to his rear. Unlike the early R-5s, the centre portion of the fuselage was skinned with detachable Alclad panels.

The Sikorsky S-51 was used by Los Angeles Airways to inaugurate the world's first scheduled helicopter service when, having been awarded a three-year mail-

Above: Los Angeles Airways used Sikorsky S-51s to inaugurate the world's first scheduled helicopter service (note US Mail painted on the forward fuselage).

Right: Westland-Sikorsky S-51 Mk 1A G-ALEI *crop spraying in Normandy in June 1949.*

carrying contract, it began services on 1 October 1947. Meanwhile, in December 1946, Westland Aircraft in England acquired the rights to construct the S-51 and market it anywhere in the world outside the U.S.A. and Canada for commercial and military use. The first example built by Westland flew for the first time on 5 October 1948 and was the first helicopter ever built by this now well-known helicopter manufacturing company. Whereas the American-built S-51 used a 450-hp Pratt & Whitney R-985 engine, Westland helicopters adopted the 520-hp Alvis Leonides 521/1 piston engine which bestowed the marginally higher maximum speed of 104 mph (167 km/h). Its range was 310 miles (499 km).

Westland Dragonflies (military S-51s) formed the equipment of the first Royal Navy helicopter squadron, No. 705, in 1950 and were operated from aircraft carriers and shore bases for air-sea rescue, ship-shore communications, planeguard and photographic duties. That same year the R.A.F. received its first Dragonflies as communications and casualty-evacuation helicopters and by April a Casualty Evacuation Flight had been formed in Malaya for jungle rescue operations during the disturbances which lasted many years. By early 1953 so much had been achieved by the Dragonfly that the Flight was enlarged to become No. 194 Squadron, the R.A.F's first fully opera-

Above: Loading mail onto a BEA S-51 for the inaugural flight of the first helicopter air mail service in the UK.

Top: Westland Dragonfly camouflaged and carrying Iraqi Air Force markings.

Left: The original Brantly B-1 co-axial twin rotor helicopter that appeared in 1946.

Opposite below: The six-passenger Doman LZ-5.

Below: Brantly B-2B, the major production version of the B-2.

tional helicopter squadron. However, Dragonflies remained in service for only three years, giving way to the much superior Whirlwind.

Meanwhile, on 1 June 1948 British European Airways (BEA) had begun the first helicopter airmail service in the U.K., flying on American-built (sometimes mistakenly said to have been a Westland-Sikorsky) S-51 over the area of Peterborough, Norwich, Great Yarmouth and Kings Lynn in England. On 14 February 1949 the same S-51 (*G-AKCU*) inaugurated the world's first experimental helicopter service by night, flying from Peterborough to Norwich. This BEA service became scheduled on 17 October but continued only until April of the following year, the inaugural flight being made by S-51 *G-AJOV* in the hands of Captain J. Cameron. Just before this, on 18 May 1949, the first helicopter base in New York became operational at Pier 41 East River. A Westland-Sikorsky S-51 was used between 9 and 19 May 1950 during the British Industries Fair, operating the first scheduled helicopter passenger services in the U.K., between London and Birmingham. From the beginning of June, BEA operated the first sustained helicopter service in the U.K. over a route between Liverpool and Cardiff, which lasted until the end of March 1951, again using Westland-Sikorsky S-51s.

A willing worker

As can be appreciated from the preceding paragraphs, many of the first postwar milestones in the use of helicopters were achieved with the Model 47 and S-51 Bell and Sikorsky went on to establish themselves at the forefront of U.S.

helicopter manufacture, with Bell concentrating on smaller single-rotor types after its flirtation with the HSL-1, and Sikorsky focussing on larger helicopters of single-rotor form. Bell has been responsible for supplying more helicopters to the U.S. forces than any other manufacturer and in early 1981 actually completed it 25,000th helicopter, a total which includes civil and military models for the home market and export. But other helicopters appeared in the U.S.A. in the 1940s, some of which helped establish their manufacturers while others faded into obscurity.

In the early postwar years Bendix Helicopter Incorporated produced some of the last U.S. helicopters to adopt a twin co-axial rotor layout, a configuration that became less and less used until Sikorsky revived the layout in very advanced form for its S-69 research helicopter of 1973. The Bendix Model J Whirlaway had a bubble-like fuselage with cabin accommodation for four persons. A similar layout was used by Newby Brantly for his B-1 two-seater, which first flew in 1946 but was considered too complicated for the civil market. Brantly went on to develop a conventional single-rotor helicopter as the two-seat B-2, which appeared in 1953. This helicopter went into production during the late 1950s and today an updated version remains in production.

Doman Helicopters experimented with a hingeless rotor in which the individual blades were dynamically flexible but unarticulated, pre-empting some of the hingeless rotor systems of the 1970s and 1980s. Doman continued work on its LZ series and in 1953 the LZ-5 appeared as a six-passenger or stretcher-carrying helicopter. Tested by the U.S.A.F. as the H-31 and CAA-certificated in 1955, it was not successful in establishing itself on the civil or military scene. However, other helicopters followed.

The Hiller XH-44, described previously, was followed in 1946 by another co-axial rotor helicopter known as the UH-4 Commuter, an open-cockpit two-seat experimental type. Further work by Hiller resulted in the experimental J-5, probably the first American helicopter to use jet thrust to counter the effect of torque. This had a stove-pipe rear fuselage, at the end of which was an orifice with vanes through which compressed air was blown. However, the J-5 was flown from August 1946 not only for the benefit of testing the anti-torque system, but also to flight-test a simplified rotor control mechanism which was adopted for the company's three-seat Model 360, its first production helicopter. In this, the two-blade main rotor was universally mounted on the shaft with a small crossing servo rotor which was connected directly to the pilot's hanging control stick. By operating the stick, the pilot altered the pitch of the servo rotor paddles which, in turn, produced aerodynamic forces to tilt the rotor head and effect cyclic pitch changes to the blades. Military and civil variants of the helicopter were built by the original company, totalling thousands of helicopters in Model 360 and UH-12 civil versions and as H-23 Raven military helicopters. In 1984 the present Hiller company, Hiller Aviation Incorporated, continues production of the UH-12, but

Above: Hiller's three-seat Model 360 photographed during certification. The servo rotor crossing the main rotor blades can be seen clearly.

Right: A US Navy Hiller HOE-1 is accompanied by an Army YH-32 with its stationary rotor showing the blade-mounted ramjets.

Opposite: The pilot of this ROE-1 Rotorcycle attaches the tail rotor boom during assembly, the complete assembly of the helicopter taking no more than five minutes.

it now has an Avco Lycoming turboshaft engine in place of the earlier Franklin piston engine.

Meanwhile, in 1950 Hiller produced its Model HJ-1 Hornet, a tiny two-seater with its two-blade rotor driven by Hiller 8RJ2B ramjet engines attached to the tips of the blades, each engine producing just 38 lb (17 kg) of thrust. An anti-torque tail rotor was not required as the rotor was self-propelling. The U.S. Navy tested three prototypes in 1951 as HOE-1s and the U.S. Army followed its initial order for two prototypes with a further order for 12 YH-32s for evaluation from 1953. All military examples were required for observation duties. However, no further production was undertaken for these armed forces. In October 1954 the Hornet was awarded a CAA Type Approval Certificate, making this helicopter the first ramjet-powered helicopter to be certificated in the U.S.A. and the 8RJ2Bs became the first tip-mounted engines for helicopters to be approved by the CAA. Hiller also produced an interesting one-man portable helicopter in the late 1950s for the U.S. Navy/Marines as the ROE-1 Rotorcycle, 12 of which were delivered. Capable of being completely folded away and then assembled and flown in five minutes or less, it was evaluated as an observation helicopter and could attain a speed of 70 mph (113 km/h). Such was its success that Saunders-Roe acquired a licence to manufacture it in the U.K. The Rotorcycle was demonstrated in all NATO countries during a four-month tour but no major production was undertaken. Gyrodyne in the U.S.A. also produced a similar machine that was tested by the Navy as the RON-1, although this had a more powerful engine and co-axial rotors. Over the ensuing years this form of ultralight helicopter has popped up now and again

Above: McCulloch MC-4, the first tandem-rotor helicopter to receive a Type Approval Certificate from the CAA.

Top: Gyrodyne RON-1 on exercise with the US Marine Corps.

for military evaluation but has never been adopted for service. One of the most recent has been the Aerospace General Mini-Copter, evaluated by the U.S. forces for more than one possible role but originally intended to be air-dropped in a container to a shot-down pilot stranded in difficult terrain or behind enemy lines so that he could make his own escape by air. All these can be said to have perpetuated the concept of ultra-light helicopters initiated by the Austrians Nagler and Baumgartl (see previous chapter).

Probably the first small helicopter with tandem rotors pylon-mounted above an enclosed cabin fuselage (for two persons) was the experimental Jov-3, development of which was entrusted in 1949 to McCulloch Motors Corporation. From the Jov-3 was produced the MC-4, which first flew on 20 March 1951 and became the first tandem-rotor helicopter to receive a Type Approval Certificate from the CAA. Nothing much came of the MC-4 but the company went on to series-produce a two-seat autogyro designated J-2 in the 1960s. Like the Jov-3, it was the work of D. K. Jovanovich.

Although Kellett had built the first U.S. helicopter with intermeshing rotors, a new company, formed in December 1945, was to become the greatest exponent of that particular layout in the U.S.A. On 15 January 1947 the Kaman Aircraft Corporation flew the two-seat experimental K-125-A, its two two-blade rotors powered by a 125-hp Lycoming engine. From this the K-190-A was developed, which was to serve as the prototype for a production derivative known as the K-225. CAA certification of the K-225 was gained in April 1949. The U.S. Navy purchased two K-225s from 1950, and this led to a contract for K-240 (Navy HTK-1) 235-hp Lycoming O-435-4 engined three-seat trainers, placed in 1950. HTK-1s were the first-ever Kaman production helicopters operated by the U.S. Navy. Meanwhile, one of the K-225s had been re-engined with a 175-shp Boeing YT50, which made this helicopter the first in the world to fly on the power of a turboshaft engine. It first flew on 10 December 1951. The U.S. Navy and U.S. Marine Corps also received, at about the same time, examples of the K-600 four-seat version, designated in service HOK and HUK respectively. A version of the Kaman K-600 for the U.S.A.F. became the well-respected H-43 Huskie, delivered from 1958 as the force's first helicopter intended for airborne fire-fighting and crash

Left: One of two Kaman K-225s acquired by the US Navy.

rescue. Although the U.S.A.F's initial version, like the U.S.N./U.S.M.C. K-600s, was powered by a large piston engine, an 860-ehp Lycoming T53-L-1A turboshaft was adopted for the H-43B version, which could carry 10 passengers or stretchers instead of the more usual fire-fighting crew and equipment.

In 1959 Kaman flew the prototype of a very different helicopter, which went into U.S. Navy service from 1962 as the H-2 Seasprite ship-based utility helicopter. Later versions took on such roles as anti-submarine and missile defence, the latter in the 1970s under the U.S. Navy's new LAMPS (Light Airborne Multi-Purpose System) programme. The SH-2F Seasprite remains in use today.

Having flown its XH-8 during the Second World War, Kellett followed this with an experimental intermeshing-rotor transport helicopter for U.S.A.F. evaluation. It was known as the XH-10. On 13 May 1954 Kellett flew the prototype of a new helicopter for the Navy, the second rotorcraft in the world to fly on rocket power. An ultralight single-seater, developed under the

Below: Just as a Kaman K-225 with a Boeing YT50 engine had become the first helicopter in the world to fly on the power of a turboshaft engine, so this Kaman HTK-1 was the first to use twin turboshaft engines.

Left: Camouflaged USAF Kaman Huskie, offering a good view of its rotor arrangement.

Opposite top: The Kaman UH-2B was the major production version of the Seasprite and was subsequently modified to other versions. U-H-2Bs eventually became Navy SH-2Ds under the LAMPS programme. Operating from anti-submarine destroyers and frigates, each SH-2D carried a Canadian Marconi search radar in a cylindrical radome under the nose and MAD (magnetic anomaly detector) gear on the starboard side of the fuselage, as seen here.

Below: Kaman Huskie, the USAF's first helicopter intended specifically for the roles of airborne fire-fighting and crash rescue.

sponsorship of the U.S. Army and Air Branch Office of Naval Research, it was built to flight-test the gyrotory stabilized control system. Power was provided by two Reaction Motors rocket motors carried at the blade tips.

Kellett also began development of the XH-17, a huge flying-crane, built under a U.S.A.F. contract of 1948, but this contract was purchased from Kellett by the Hughes Aircraft Company in August that year to launch Hughes onto the helicopter scene. The Hughes XH-17 was really a testbed aircraft but nevertheless was a most remarkable helicopter. With a gross weight of 43,000 lb (19,504 kg), well over twice that of any contemporary helicopter, it made its maiden flight on 23 October 1952. Two modified General Electric GE 5500 turbojet engines were used to supply gas pressure through ducts to outlets near the tips of the 130-foot (37.62-metre) diameter rotor. The dual hydraulic system of the XH-17 was similar to that used in the gargantuan Hughes H4, the world's largest flying-boat of all time. By adopting very long legs, the XH-17 was able to straddle bulky cargoes before attachment and then lift-off. The U.S.A.F. had intended to follow-up the XH-17 with the refined XH-28 but financial restrictions as a result of the outbreak of the Korean War prevented this.

In October 1956 the prototype of the first commercially successful Hughes helicopter appeared and flew as the Model 269. Five examples of this light two-seater were evaluated by the U.S. Army as observation helicopters. Between 1964 and 1969 more than 790 were delivered to the Army as TH-55A Osage primary helicopter trainers. The version for the civil market became the Model 269A and some were also exported. From the Model 269/269A was developed the Hughes Model 300, which remains in production.

Another notable giant, which actually preceded the XH-17, was McDonnell's XHJD-1 Whirlaway. Designed and built to a U.S. Navy contract of 15 May 1944, it clearly owed a lot to the Platt-Le Page. It was first flown on 27 April 1946 when, as a six-ton machine, it was undoubtedly the world's heaviest and largest helicopter. Flying continued until about 1949, during which time it spent more than 250 hours in the air but no production followed. Similar fates attended the XH-20 Little Henry of 1947, a tiny helicopter with ramjet compressors

Above: Operational equipment assigned to LAMPS Seasprites included two Mk 44 or Mk 46 torpedoes and fifteen active and passive sonobuoys (nine seen here) fired from the fuselage side by explosive charge.

Below: The RotorCraft RH-1 Pinwheel was the first rocket-powered helicopter to fly, in April 1954. Each rotor-tip motor weighed less than 1 lb and used 90 per cent hydrogen peroxide as a mono-propellent.

Right: Hughes TH-55A Osage primary flight training helicopter.

Left: When tested, the Kellet XR-10 (later XH-10) was the USAF's largest helicopter, and the first all-metal helicopter in the world. Accommodating ten troops, it could fly at more than 100 mph (161 km/h).

Below right: Hughes Model 300C equipped for agricultural crop spraying.

Below: After working on the XH-17 and XH-28 programmes, Hughes continued research of pressure jet systems. One outcome was the hot-cycle system, in which the efflux from two General Electric YT64-GE-6 gas generators was ducted to blade-tip cascades to turn the rotor. To flight test the concept, the XV-9A was built, first flown in November 1964. Though reliable, the hot-cycle system proved too heavy for wider applications.

driving the rotor, and the small Model 120 flying-crane or 12-passenger (in detachable pod) helicopter of 1957 which McDonnell developed with its own funding. The Model 120 featured pressure-jet rotor drive. Another interesting McDonnell aircraft of this period was the XV-1, an experimental convertiplane that, in 1956, set an unofficial speed record for helicopters at 200 mph (322 km/h).

In the year McDonnell flew its Little Henry, 1947, the first American helicopter to feature all-metal rotor blades appeared as the Sikorsky S-52. First flown on 12 February 1947, it received CAA approval for both day and night commercial operation in December that year and in 1948 established international records for speed and height. The four-seat S-52-2 version was evaluated by the U.S.A.F. as the H-18 but was not adopted for operational use. The U.S. Marine Corps, however, received a considerable number for service from 1952 and a few also served with the U.S. Coast Guard.

In November 1949 Sikorsky opened a new era in helicopter design, when the first prototype of its S-55 made its maiden flight at Bridgeport. One of five YH-19s ordered for trials with the U.S.A.F., the unique feature of the helicopter was the location of its large 550-hp Pratt & Whitney R-1340 engine; it was carried in the nose on an angular mounting with a drive shaft that sloped to the base of the rotor pylon. By adopting this layout, the main cabin, situated directly under the main rotor, remained completely clear.

The U.S.A.F. received H-19A and H-19B versions, the latter the more powerful version with a 700-hp Wright R-1300-3 engine and capable of accommodating 10 troops, the equivalent amount of freight or eight stretchers for rescue work. The U.S. Army received similar H-19C and H-19D Chickasaws for its Field Forces. The H-19B had a maximum speed of 112 mph (180 km/h).

The first of the U.S. Navy versions was the HO4S-1. It was similar to the U.S.A.F.'s H-19A but was used from 1951 in small

Above: The huge McDonnell XHJD-1 Whirlaway dwarfs the tiny XH-20 Little Henry.

Right: The first helicopter with all-metal rotor blades was the two-seat Sikorsky S-52.

Opposite top: The helicopter proved an ideal platform from which to spray hillside crops. Here a Hughes Model 300C takes the Mozelle Valley vineyards in its stride.

Opposite bottom: The Hughes Model 300C has also been built in Italy by BredaNardi.

Left: USAF Sikorsky H-19B Chickasaw operated by MATS for air rescue.

Opposite top: Westland Whirlwind Series 1 in commercial service with Bristow Helicopters.

Opposite below: Using Sikorsky S-55s, Sabena inaugurated the world's first international helicopter services in September 1953. Here a Sabena S-55 takes off from Brussels city centre heliport.

Below: Early military Westland Whirlwinds supplied to Austria.

numbers as a general purpose transport with squadron HU-2. The next Navy version was the HO4S-3, which was similar to the H-19B but was intended for anti-submarine duties. It formed the equipment, initially, of squadron HS-1, the U.S. Navy's first ASW helicopter squadron that was commissioned at Key West in Florida on 3 October 1951. The U.S. Coast Guard received HO4S-3Gs for rescue work, while the U.S. Marine Corps received HRS assault transports.

The incredible success and versatility of the S-55 led to licence-production being undertaken in the U.K. by Westland as the Whirlwind, by SNCASE in France as the Joyeux Eléphant and by Mitsubishi in Japan. Total production in the U.S.A. by Sikorsky in both military and S-55 civil forms was 1,281 examples, while Westland produced over 400 Whirlwinds and small numbers were completed in Japan and France. These entered military and/or commercial service in a great many countries through exports. Westland Whirlwinds, flown from 1952, underwent the greatest development and included Series 2 aircraft with 750-hp Alvis Leonides Major engines and Series 3 with 1,050-shp Rolls-Royce Bristol Gnome H.1000 turboshaft engines. Turboshaft power was fitted to some Sikorsky S-55s when, in the 1970s, Aviation Specialties began modifying S-55s into S-55Ts, which they did with 840-shp Garrett-AiResearch TSE 331-3U-303 turboshaft engines that had been derated to 650-shp.

Whirlwinds were delivered to the Royal Navy, R.A.F. and British Army as well as commercial and export customers. In Navy service the Whirlwinds followed an earlier batch of 25 U.S. Sikorsky S-55s that had been accepted under the Mutual Defense Assistance Program. Some of these joined No. 848 Squadron, the first Royal Navy helicopter squadron ever to go into action when, in March 1953, they undertook their first operations in Malaya. American-built S-55s also equipped No. 845 Squadron, the

Royal Navy's first anti-submarine helicopter squadron.

The S-55's impact upon military and commercial helicopter operations is incalculable: a very similar helicopter appeared in the Soviet Union in 1952; more details later. Between 13 and 31 July 1952 two Sikorsky S-55s completed, in stages, the first-ever west-east helicopter crossing of the North Atlantic. On 1 September 1953 the Belgian airline Sabena flew S-55s to inaugurate the world's first international helicopter services, flying from Brussels to cities in the Netherlands and France as well as to Maastricht. These are only a few of the many milestones achieved by the S-55.

Larger by design

The first large Sikorsky helicopter and the first Sikorsky with twin engines was the S-56, developed originally to meet a requirement of the U.S. Marine Corps for an assault transport helicopter accommodating 20 troops or 24 stretchers or cargo (including vehicles) loaded through the clamshell nose doors and side door. The prototype S-56 flew for the first time on 18 December 1953 and delivery of HR2S-1s to U.S.M.C. units began in 1956. Two were modified in 1957 into HR2S-1W early-warning helicopters, each with an AN/APS-20E radar scanner housed in a large undernose radome. As the largest helicopter in existence outside the Soviet Union, it was also ideal for U.S. Army service and this force received the greatest number as H-37A Mojave transports, each accommodating 23 troops or stretchers or cargo. In 1956 an HR2S-1 set a new world speed record of 162.67 mph (261.8 km/h) and two altitude records with payload. In one record it lifted 6,000 kg to an altitude of 7,000 feet (2,134 metres). In the U.K., Westland considered licenced production but

Above: Prototype Sikorsky S-58.

Top: One of the first Sikorsky H-37A Mojaves delivered to Army Aviation.

Opposite top: By flying from Athens, Greece, in a USMC HSS Seahorse helicopter to the deck of the cruiser USS Des Moines, *then sailing in the Mediterranean, President Eisenhower became the first US President to land by aircraft on board a ship. Navy HSS and Marine HUS helicopters were redesignated H-34 under the standardized tri-service military designation system of 1962.*

Left: US Marine Seahorse and Mojave helicopters on board a US Navy carrier.

Right: The turbine-powered (Gnome) Westland Wessex HC.Mk 2 transport, ambulance and general-purpose helicopter for the RAF entered service in 1964 as a high-performance development of the Royal Navy's Gazelle-powered HAS.Mk 1 ASW helicopter.

rejected the idea in favour of producing its own large helicopter.

Whilst the S-56 remained only a military helicopter in the U.S.A., Sikorsky's follow-up to the S-55, the larger S-58, became an internationally best-selling helicopter in both military and commercial forms. First flown as a prototype on 8 March 1954, it joined the U.S. Navy under the designation HSS Seabat, the U.S.M.C. as the HUS Seahorse and the U.S. Army as the H-34 Choctaw. Capable of 123 mph (198 km/h) on the power of a 1,525-hp Wright R-1820 piston engine, the helicopter had accommodation which varied according to its roles the latter could include anti-submarine and search and rescue. In the assault configuration, accommodation for 18 troops was normal.

Apart from exports, S-58s were built in France by Sud-Aviation, in Japan by Mitsubishi and a turboshaft-engined version was produced in the U.K. by Westland. The latter, named Wessex, was first fitted with a 1,450-shp Napier Gazelle 161 turboshaft engine and in this form it joined the Royal Navy as an anti-submarine helicopter in 1961. The first version

with coupled Rolls-Royce Gnome turboshafts flew in January of the following year and this power plant became standard thereafter. Apart from Royal Navy and R.A.F. service, military exports were made to Iraq, Ghana and Brunei, and the Wessex also entered commercial operation.

Before Sikorsky began production of the helicopters familiar today, like the S-61 and S-62, it made its first attempt at a large specifically designed flying-crane helicopter designated the S-60. Using components of the S-56, it had a tail-boom type fuselage aft of the cockpit under which cargo or vehicles could be suspended for hauling or a detachable pod could be attached in which 20 troops or other equivalent loads could be accommodated. Although the S-60 was not taken up by the U.S. Army after evaluation, it provided a worthwhile step towards Sikorsky's later production flying-crane, the S-64 Skycrane of 1962.

As mentioned in the previous chapter, the U.S. Piasecki HRP-1 became the first successful twin-rotor helicopter to enter production and service. Understandably receiving the nickname 'Flying Banana', the 54-foot (16.46-metre) long helicopter could attain 104 mph (167 km/h) on the power of a single 600-hp Pratt & Whitney engine. Accommodating a crew of two in tandem and eight passengers or six stretchers, 20 went to the U.S. Navy unit VX-3 and U.S.M.C unit HMX-1 under the name Rescuer and a small number

Above: As a ship and shore based assault helicopter, Westland built the HU.Mk 5 for the Marine Commandos. This version could also be armed for attacking ground and sea targets. The Wessex (illustrated) is carrying homing torpedoes, 2-in (5-cm) rockets in pods and Nord SS.11 wire-guided anti-tank missiles.

Opposite top: Sikorsky S-60 prototype flying-crane helicopter carrying as a slung load a platform on which is secured fire-fighting gear and a cement mixer. Powered by two Pratt & Whitney R-2800 engines, it was evaluated by the US Army during August and September 1959.

Right: Piasecki HRP-1 'Flying Banana' demonstrating its ability to hover while Navy personnel climb on board.

were also operated by the U.S. Coast Guard. The rather unpleasant lines of the HRP-1 were refined with the follow-on HRP-2 which, apart from configurational changes to reduce drag and the adoption of all-metal fuselage skins instead of the HRP-1's forward fabric covering, had a more rounded nose with the two crew members seated side by side. HRP-2s were delivered to the U.S.M.C. for developing the techniques of airborne amphibious assault.

A more powerful version of the HRP-2 was ordered for the U.S.A.F. and U.S. Army as the H-21, versions of which also found favour with the air forces of Canada and West Germany and the French Navy. First flown in April 1952, H-21As with 1,150-hp Wright R-1820-103 engines were operated by the U.S.A.F. as Arctic rescue helicopters and accommodated 14 troops, 12 stretchers or freight. The H-21B, of which the U.S.A.F.

received no fewer than 163, used the same engine as the H-21A but in the H-21B it was derated to the higher level of 1,425 hp. This version was capable of carrying six extra troops and had the option of removable armour to protect vital components and self-sealing fuel tanks. It equipped Troop Carrier Command from 1953.

The H-21C was, however, the major version of the helicopter, and more than 330 were built from 1953. The 'C' served as a U.S. Army transport for freight or troops and was named the Shawnee in service. (The name Workhorse was given to the other helicopters of the H-21 series.) While in production, the H-21C witnessed the metamorphosis of the Piasecki Helicopter Corporation into the Vertol Aircraft Corporation in 1956. A commercial version of the H-21 was the Model 44, that gained its CAA certific-

Above: USAF H-21C employed as a rescue helicopter using pontoons.

Top: German Army Workhorses used for utility transport duties.

Opposite top: HRP-1s gave the US forces the ability to train in new forms of warfare. These included air-dropping large numbers of assault troops into combat areas and demonstrating new minesweeping techniques.

Opposite bottom: Five HRP-2s were delivered to the US Marine Corps.

ation in 1957. Recipients included such non-civil operators as the French Government, the R.C.A.F., the Japan Ground-Defence Force and the Swedish Navy, the latter using it for various duties including anti-submarine. As with earlier Piasecki helicopters, the normal wheel undercarriage could be replaced by pontoons for operation from water, although Model 44s supplied to the Swedish Navy and New York Airways had water-tight lower fuselages and triple rubberized-fabric floats around the undercarriage legs for use in an emergency ditching. It is of interest to note that R.C.A.F. helicopters were operated by Spartan Air Services to carry supplies to the Mid-Canada chain of early-warning stations. Accommodation varied according to the version and role, but the Model 44A could seat 19 civilian passengers, 20 troops, 12 stretchers or 2½-tons of freight.

The H-21 was the U.S.A.F.'s first tandem-rotor helicopter in service. Even before the prototype flew, the U.S.A.F. had placed an order for a large quantity of H-25As for use by the U.S. Army. These were five-troop or three-stretcher helicopters of typical Piasecki tandem-rotor layout but smaller and neater. In Army service they were named Army Mules, and deliveries took place between 1953 and 1954.

Like the H-21, the H-25A was a version of a Navy helicopter, this time the HUP-2 Retriever, deliveries of which had been underway since 1949 in HUP-1 form without autopilot, in HUP-2 form with modified Sperry autopilot, and in HUP-2S form. Except for the HUP-2S, which was specifically an interim anti-submarine warfare helicopter with sonar equipment, the HUPs were intended for a variety of duties which included ship-board plane guard, rescue, observation and intership and ship-to-shore utility operations. Because it was required that this helicopter be capable of using the smallest lift on board an aircraft carrier without having its rotor-blades folded or of using the aircraft lift on a cruiser with blades folded, it was the smallest Piasecki tandem-rotor helicopter. In addition, the Navy later received most of the Army's H-25As as HUP-3s for cargo carrying and medical evacuation.

Piasecki's largest helicopter was the giant PV-15 Transporter. Two prototypes were tested in the mid-1950s by the U.S.A.F. as 40-troop, 32-stretcher or cargo transports. Powered by two 3,000-shp Allison T38 turboshaft engines in YH-16A form (the first prototype, the YH-16, had Pratt & Whitney piston engines installed), it was capable of a respectable 125 mph (201 km/h) but was not selected for further production.

Other Vertol tandem-rotor helicopters followed, but in March 1960 Vertol became a new division of the Boeing Company. Boeing Vertol products are mentioned in a later chapter but it is interesting to note here the close configurational link between the HUP/H-25 and the subsequent Sea Knight and Chinook, both of which were already under development when Boeing took over Vertol. Only one new aircraft produced by Vertol

Bottom: US Navy Piasecki HUP-2 Retriever, the first production helicopter to be equipped with an autopilot to allow 'hands off' flying.

Below: New York Airways operated the Vertol Model 44 with special safety features, which included a water-tight lower fuselage and triple rubberized-fabric floats in case of an emergency ditching. This NYA Vertol 44 lifts-off from the West 30th Street Heliport in 1958.

Above: The uniquely configured Vertol Model 76, which performed 448 flights including 239 partial and 34 full transitions from vertical to horizontal flight.

Left: The world's largest tandem-rotor helicopter at the time of its appearance, the 15-ton Piasecki PV-15 Transporter is seen during its first flight on 23 October 1953.

carried out most of its flying before the formation of Boeing Vertol. This was the Model 76 (Army-Navy VZ-2A). Intended for research only, it used a tilting wing carrying rotor/propellers to achieve vertical, transitional and horizontal flight. The VZ-2A made its maiden flight in August 1957.

All shapes and sizes

Prior to Piasecki Helicopter Corporation becoming Vertol Aircraft Corporation, the Chairman and President of the former company had, in 1955, set up Piasecki Aircraft Corporation. The products of this concern included a unique series of vertical take-off aircraft *cum* land cars tested from 1958 under the name Airgeep. With either piston or, later, turboshaft engines powering opposite rotating ducted rotors, the Airgeep was intended mainly for observation with the crew making use of natural cover to avoid detection but with the option of driving on roads. Though successful in trials, the Airgeep and its variants were not adopted for service.

Earlier paragraphs have outlined man-portable ultralight helicopters which were evaluated by the U.S. military. Another very lightweight helicopter of the 1950s was the American Helicopter XH-26, a single-seater with a two-blade rotor driven by tip-mounted pulse-jet engines. Although it did not win any production orders from the U.S. Army, it was particularly inter-

Bottom: The Piasecki Airgeep II was an improved version of the original Airgeep and was first flown in 1962. It used Turboméca Artouste IIC turboshaft engines to drive the two three-blade ducted rotors and had powered wheels for ground mobility when the terrain allowed.

Below: The first American Helicopter XH-26, using XPJ49-AH-3 pulse-jet engines.

esting in that it had a fully enclosed cabin for the pilot and yet was capable of being collapsed to fit inside a container no larger than 5 by 5 by 14 feet (1.52 by 1.52 by 4.27 metres) for air-dropping into a combat zone. Two men could assemble it and have it airborne in approximately 20 minutes.

A name not readily associated with helicopters is that of Cessna. However, when Cessna acquired the Seibel Helicopter Company, it broke into the helicopter market temporarily. (Seibel, before the takeover, had produced a single example of a rather peculiar helicopter with separate canvas enclosures for the pilot and passenger/cargo, which had been tested as the YH-24.) The Cessna CH-1, which first flew in July 1954, had two notable features. The most obvious was the installation of the Continental piston engine in the helicopter's nose, while the most important was the use of flexible L-section hinges to attach the rotor blades to the rotor hub. These hinges allowed the pitch of the blades to be varied without the usual pitch-change bearings. Ten were evaluated by the U.S. Army as four-seat YH-41 Senecas, and 30 CH-1C Skyhook civil counterparts were constructed up to 1960.

The preceding paragraphs have detailed some fairly unusual helicopters but none matched the de Lackner DH-4 Aerocycle, which was ordered for evaluation by the

Above: Cessna's venture into helicopter construction produced the CH-1 Skyhook.

Top: de Lackner DH-4 Aerocycle on test at Camp Kilner during late 1955.

Above: Bristol Sycamore 3s operated by the Royal Australian Navy.

Top: Bristol Sycamore HC.14, once used for rescue missions but now destined for display purposes only.

U.S. Army. First flown in January 1955, the DH-4 comprised a small circular platform on which was carried a two-stroke engine and handle-bar controls (the hollow control column also acting as a 1 U.S. gallon fuel tank). The pilot stood upright on the platform while below his feet turned two contra-rotating rotors of 15 feet (4.6 metres) in diameter. The helicopter rested on water via five air bags. Despite its strange appearance, it proved stable in flight, was capable of a speed of around 75 mph (121 km/h), could hover at high altitude and land by autorotation with power off, and could lift light cargo. U.S. Army personnel proved able to master the DH-4 in approximately 20 minutes. This was by no means the last flying platform developed for, and tested by, the U.S. forces. Later machines included the turbofan-engined Williams WASP of the 1970s.

Postwar developments in Britain

After the Second World War, Westland Aircraft and the Bristol Aeroplane Company quickly emerged as the leaders of British helicopter manufacturing. But, whereas Westland achieved remarkable results from developing and constructing Anglicized versions of

U.S. Sikorskys, Bristol designed its own helicopters and so historically must rank as the more important of the two.

Bristol, one of Britain's longest established aircraft manufacturers, was founded in 1910. After the Second World War it was responsible for developing the Britannia, the first turboprop airliner to go into service with BOAC (British Overseas Airways Corporation, Britain's national airline). The company's first rotary-winged aircraft was the Type 171 Sycamore, which first took to the air on 24 July 1947. In original form the Sycamore was intended as a two-crew and four-passenger air-taxi and feeder helicopter. Thus July 1947 also marked the date of the first British commercial helicopter to fly. Naturally the Sycamore was also the subject of the first Certificate of Airworthiness awarded to a British helicopter.

In layout the Sycamore was a typical single main rotor and anti-torque tail rotor helicopter. Three prototypes were built: two had the same type of Pratt & Whitney R-985 engine that powered a number of successful U.S. helicopters; the third, however, had a more powerful 525-hp Alvis engine installed and an Alvis Leonides thereafter became the standard for production, which began with the civil Type 171 Mk 3. In 1951 the British Army began receiving Sycamores for ambulance, communications, observation and transport roles under HC.Mk 10 and Mk 11 designations. In the early part of the following year the R.A.F.'s Coastal Command received HR.Mk 12 search and rescue, reconnaissance and communications helicopters. These were followed, from 1953, by HR.Mk 13s and Mk 14s for search and rescue, communications, cargo, VIP transport, and ambulance duties with Fighter Command. The first Sycamore squadron of Fighter Command was No. 275, which had the honour of becoming the R.A.F.'s first operational helicopter squadron in the U.K. for search and rescue.

R.A.F. Sycamores entered the fray in Malaya in 1954 and were used to develop the technique of lifting troops to positions in mountainous areas that no other form of transport could reach. Other Sycamore operators included the Royal Australian Navy and the Federal German Air Force and Navy. Its maximum speed was 127 mph (204 km/h).

Bristol was also responsible for the first British-designed helicopter with twin engines and twin rotors when, on 3 January 1952, the prototype Type 173 made its maiden flight at Filton as the civil registered *G-ALBN*. Each Alvis Leonides drove a four-blade rotor carried on top of pylons above the two-crew and 13-passenger cabins. In the following year it landed on board the Royal Navy aircraft carrier HMS *Eagle*, by which time it carried military markings.

The second Type 173 prototype was similar to the first, except when temporarily fitted with auxiliary wings to gain data on the effects of these on forward flight performance. In July 1956 one was handed over to BEA for handling trials and thus became the first twin-engined helicopter to be delivered to an airline. More powerful

The first aircraft carrier trials by a twin-engined helicopter were conducted on HMS Eagle *using the prototype Bristol Type 173.*

Above: RAF Westland Belvedere HC.Mk 1 transporting a Bloodhound anti-aircraft missile.

Right: Belvedere HC.Mk 1 helping to place a structure on the roof of a high building.

versions of the Type 173 were completed for research and development but the helicopter did not enter production, largely as the result of engine development problems. From the Type 173 Bristol designed the similar Type 191 for the Royal Navy, the Type 192 for the R.A.F. and the Type 193 for the Royal Canadian Navy. Only the Type 192 with its two 1,650-hp Napier Gazelle NGa.2 Mk 101s was built. It was capable of 138 mph (222 km/h) and accommodated 19 troops, up to 30 survivors in an emergency casualty evacuation role, 12 stretchers or 6,000 lb (2,722 kg) of internal or external freight.

Twenty-four Type 192s were built for the R.A.F. : the first flew on 5 July 1958, and this date marked the first flight date of any Type 192, as no prototypes were deemed necessary. However, on 23 March 1960, Westland Aircraft took over the helicopter division of Bristol and the Type 192s became known as Westland Belvedere HC.Mk 1s. The first R.A.F. Belvedere Squadron was No. 66 in 1961 and this squadron, one of three that operated the type, was also the last in 1969.

Westland's own activities during the 1950s produced the company's first helicopters of its own design. By installing an Alvis Leonides in its licence-built Sikorsky S-51s, Westland had produced the Dragonfly which offered a considerable increase in engine power and therefore better payload. However, the Dragonfly's fuselage was not ideal for making the best use of increased passenger/cargo capability and so Westland designed its Widgeon. Apart from its more modern fuselage with seating for five persons (including the pilot), the Widgeon adopted the improved rotor of the Whirlwind helicopter (Sikorsky S-55) which had larger rotor blades attached to the rotor hub via offset flapping hinges. Unlike the Dragonfly's free-flapping blades on centreline hinges, the offset hinges offered improved efficiency and a greater centre-of-gravity range. However, in other respects the Widgeon *was* a Dragonfly and, indeed, was known as the S-51 Series 2.

Like other helicopters of the period, the Widgeon was capable of land or water operation (using pontoons) and could be fitted out for a variety of roles in addition to passenger carrying, such as ambulance and search and rescue. As an aerial crane, it could lift 1,000 lb (454 kg). However, for Westland the timing was not in its favour. Westland considered the smaller Widgeon would complement the higher performing and larger capacity Whirlwind, but, in fact,

Westland Widgeon, a direct development of the Dragonfly.

the Whirlwind proved so popular that no market could be found for the Widgeon. Only 14 Widgeons were built; some, like the first, were conversions of Dragonflies.

As mentioned earlier, Westland considered the production of the Sikorsky S-56, as the company had already built up a strong manufacturing link with the U.S. company. This plan, however, was finally dropped when an alternative idea was accepted. Westland decided to develop its own helicopter of even larger proportions than the S-56, capable of undertaking military roles and able to accommodate 45 passengers.

The resulting Westland Westminster was the most ambitious helicopter project undertaken by the company but one which had to stay within strict limits of cost and staff involvement as the Wessex programme was also then active. As a result, the first prototype was flown as a test rig, with the majority of the fuselage structure uncovered and weights added to bring the helicopter up to the anticipated all-up weight of a production Westminster. The main innovation was the choice of two 3,150-shp Napier Eland E229A turboshaft engines carried above the fuselage structure to drive an S-56 rotor via S-56 reduction gear. The tail rotor was also of S-56 type. The use of Sikorsky components helped to keep development costs low. In its selection of turboshaft engines, Westland was one of the first aircraft manufacturing companies to appreciate the benefits that these fairly light-weight but high-rating engines bestowed, especially when carried above the fuselage to render the maximum cabin area for passenger/cargo carrying.

The original Westminster prototype flew for the first time on 15 June 1958. This was later skinned with fabric to expand development trials while the properly skinned second prototype was being built. The new prototype appeared in the summer of 1959 but the complete programme was abandoned in 1960 after Westland decided that its newly acquired Bristol projects were likely to be better.

In August 1959 Saunders-Roe was taken over by Westland, followed in 1960 by Fairey Aviation. Saunders-Roe had been developing a compact five-seat general-purpose helicopter with turboshaft power (Blackburn) as the P.531; it had first flown on 20 July 1958. Intended for civil and military use, one of its projected

Above: British Army Westland Scout AH.Mk 1s occupied in controlled flying demonstrations.

Opposite: Widgeon G-ANLW constructed from an S-51 and using neoprene pontoons.

Westland Westminster (closest) flying alongside a Royal Navy Wessex, a Whirlwind and a Widgeon.

Above: The Ugandan Police took delivery of two Westland Scouts (as seen), as did the police department of Bahrain.

Right: The Cierva two-seater was produced by Saunders-Roe as the Skeeter, powered by a 200-hp de Havilland Gipsy Major engine. This Skeeter was operated by the British Army.

military roles involved carrying weapons. From the P.531, Westland developed the production Scout, a land-based general-purpose helicopter, adopted for service with the British Army from 1963 and exported in small numbers, and the Wasp anti-submarine helicopter that joined the Royal Navy the same year for operation from frigates. The Wasp was also exported. British Scouts and Wasps remained in use until they were superseded by Lynx helicopters.

The P.531 itself had been basically a larger development of the Skeeter, a two-seat light piston-engined helicopter put into production by Saunders-Roe for the British Army as an observation type, and in very small numbers had been supplied to the R.A.F. as a CFS trainer. It was also exported to the Federal German Navy and Army.

However, Saunders-Roe had not originated the Skeeter but had acquired it through the purchase of the Cierva Autogiro Company in January 1951. As a Cierva helicopter, it had first flown on 8 October 1948.

The Skeeter had not been the only Cierva helicopter of the postwar years, as the company had successfully made the transition from autogyros to helicopters at an earlier date. Worthy of mention is the Cierva W.9 of 1947, an experimental helicopter that used jet thrust ejected from the tail to

Royal Navy Westland Wasp HAS.Mk 1, a type of helicopter ordered for anti-submarine duties from Leander, Rothesay *and* Tribal *class of Royal Navy frigates, but not possessing search radar. Unlike the similar Scout, the Wasp used a quadricycle castoring wheel undercarriage and had a more powerful engine.*

Right: Cierva Air Horse, its structure dominated by the three rotors.

Below: The two-seat Cierva W.9 used jet thrust to counter torque.

counter torque effect. Actual power for the W.9 was from a 200-hp de Havilland Gipsy Six Series III piston engine, the jet efflux being a mixture of exhaust gases and air.

Historically far more important than the W.9, although the W.9 was the first British helicopter with jet thrust for anti-torque, was the Cierva Air Horse. With great foresight, Pest Control Limited approached Cierva with the prospect of acquiring a large helicopter for its crop-spraying operations. A gross weight of 18,000 lb (8,165 kg) was quoted. This directive was almost certainly the first to which a crop-spraying helicopter was specifically designed and flown. Because of the very heavy weight requirements and the restriction on the diameter of any rotor to about 50 feet (15 metres) given the known state of the art, the W.10 was designed to use three three-blade rotors powered by a 1,620-hp Rolls-Royce Merlin 24 engine. The effect of the three rotors was that the whole structure was dominated by heavy outriggers from the fuselage sides to carry two rotors and a single boom over the crew cockpit to mount the third. Also envisaged by the company were the W.11 24-passenger transport and the W.12 12-passenger transport, the engines and fuselages of these later designs varying (the W.12 with two Alvis Leonides). An important feature of the design layout was the angle of tilt in the planes of rotation to counter torque effect, as all three rotors turned in the same direction.

Two Air Horse prototypes were ordered by the Ministry of Supply, the first making its maiden flight in December 1948. It was then the largest single-engined helicopter in the world. The structures for both helicopters were built by Saunders-Roe. By October 1949 the first had completed its initial trials up to its gross weight of 17,500 lb (7,938 kg) but this crashed in mid-1950 when a rotor hub failed. The second Air Horse also suffered rotor failure and the whole programme was then abandoned. Pest Control, however, continued their helicopter aerial crop-spraying experiments with a Westland-Sikorsky S-51.

In a previous chapter the Fairey Rotodyne was detailed which used pressure-jets to power the main rotor. Two years before the Rotodyne flew, Fairey had used a pressure-jet system to power the two-blade rotor of a diminutive two-seat helicopter known simply as the Fairey Ultra Light Helicopter. With compressed air fed to rotor-tip units by a Blackburn-built Turboméca Palouste turbo-compressor, this helicopter made its maiden flight on 14 August 1955 and was, therefore, the first British helicopter to feature pressure-jets. It was intended primarily as an Army observation and liaison helicopter, but only the prototypes ordered by the Ministry of Supply were built.

The first Fairey Ultra Light Helicopter prototype.

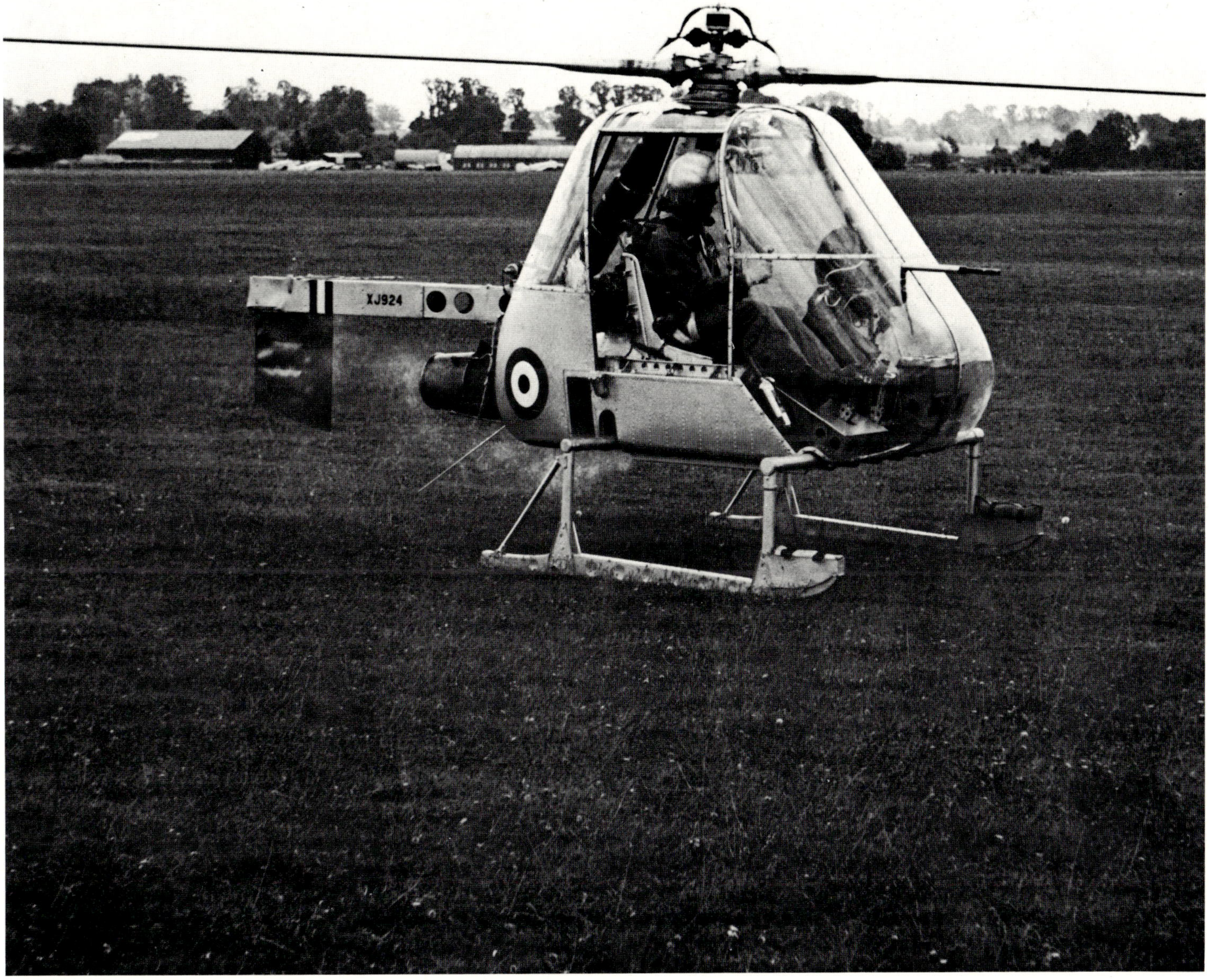

Right: Sud-Ouest S.O.1120 Ariel III, first flown on 18 April 1951, and with its turbine engine-compressor exposed for early trials.

Opposite: An Alouette II equipped with Nord SS.11 wire-guided anti-tank missiles, one of the first helicopters to be armed with anti-armour missiles.

Below: A small French S.O.1221 Djinn helicopter equipped for agricultural spraying on display at an air show, contrasting dramatically with the large Soviet Mi-6 behind.

A slow beginning for France

Although France had been at the forefront of helicopter technology until the outbreak of the Second World War, the nation was slow in getting rotorcraft into production after the war. One or two French helicopters have already been mentioned, including an example of the German Fa 223, but most helicopters that appeared in the first postwar decade remained experimental. A landmark in the fortunes of the French helicopter industry appears to have been the formation of Sud-Aviation in March 1957 from the former Ouest-Aviation (SNCASO) and Sud-Est Aviation (SNCASE).

Not surprisingly Breguet got off to an early start, producing a series of helicopters with fully enclosed cabins and twin co-axial rotors. The company's Gyroplane III accommodated a pilot and four passengers and was powered by the internationally favoured Pratt & Whitney Wasp engine. But this remained a prototype and thereafter Breguet concentrated on fixed-wing aircraft for commercial and military customers. In the late 1940s Aérocentre designed a very attractive five-seater for commercial operation as the N.C.2001 Abëille, which featured intermeshing twin rotors for the first time on a French helicopter. Nothing came of this. About the same time Nord experimented with its 1700 Norélic, which featured a large tail fan carried pusher fashion and surrounded by a short-chord ring. Torque correction and directional control were achieved by using this fan, its slipstream blowing onto vertical and horizontal control surfaces. While the fan was recognized to be less efficient than a conventional anti-torque rotor, Nord thought this less important than the benefits of simplicity, safety and the elimination of vibration. Another benefit was that the tail fan allowed the Norélic to be manoeuvred on the ground under its own power without the main rotor turning.

Sud-Ouest produced a series of small helicopters between 1947 and 1951 in the Ariel series. Each had a single main rotor, the blades of which had tip-jets to burn air supplied from a compressor and fuel. The same concern also built the S.O.1310 Farfadet, the first French convertiplane. It was able to operate as a helicopter using the tip-jets of its Ariel-type main rotor or as an autogyro for horizontal cruising flight using its nose-mounted Turboméca Artouste II

turboprop engine and propeller and with extra lift provided by short wings. First flown on 8 May 1953, the Farfadet achieved its first transition from vertical to horizontal flights on 1 July 1953 but this five-seater remained experimental.

Sud-Ouest did manage to put into production a helicopter with a tip-jet rotor system as the S.O.1221 Djinn, which followed the single-seat S.O.1220. Indeed, the Djinn became the world's only helicopter with a pressure-jet system to go into quantity production. Unlike on the Ariels and Farfadet, the rotor-tip ejectors were not combustion chambers as no fuel was burned. Instead the rotor was driven only by supplies of compressed air from the 240-hp Turboméca Palouste IV turbo-generator.

The first prototype S.O.1221 Djinn two-seater made its maiden flight on 16 December 1953, followed by a pre-production example in September 1954 and the first production model on 5 January 1956. A Certificate of Airworthiness was granted in France in May 1957 and a Type Approval Certificate from the U.S.A. was awarded in April 1958. Intended for commercial and military operations, its optional equipment included two externally mounted stretchers and agricultural crop-spraying equipment. Capable of 78 mph (125 km/h), 150 production helicopters were built. The French Aviation Légère de l'Armée de Terre received 100, which were used for observation, liaison, stretcher-carrying and training, and the type also took part in launching trials of the Nord SS.10 wire-guided anti-tank missile. Three were evaluated by the U.S. Army as HO-1s and six went to the Federal German Defence Ministry. The others went to civil customers in several countries.

The most important early helicopter of French origin was undoubtedly the Sud-Est S.E.3130 Alouette II; the first prototype made its maiden flight on 12 March 1955. The story of the Alouette really began in 1948 when, along with the French-built example of the German Fa 223, Sud-Est completed the single-seat S.E.3101. Designed more on the lines of the early U.S. Sikorskys, it was a single-seat helicopter powered by an 85-hp Mathis G4R four-cylinder engine. Its most unusual feature was its twin three-blade anti-torque tail rotors carried on 'butterfly' surfaces above the tail of the open-structure fuselage. This experimental helicopter proved capable of 80 mph (129 km/h).

From the S.E.3101, Sud-Est moved on to its S.E.3110, intended as a two-seat civil helicopter. It retained a rotor layout similar to the S.E.3101, was powered by a 200-hp Salmson 9 NC radial engine and had an egg-shaped cabin. Next in the line of prototypes was the S.E.3120, powered by a 200-hp Salmson 9 NH engine but now with a new cockpit for the two crew members, an open-structure rear fuselage and a single two-blade tail rotor. Intended for agricultural spraying and dusting but with provision for two stretchers, the first of two prototypes took to the air on 31 July 1951 and received the name Alouette.

Production of the helicopter began with the S.E.3130 Alouette II, a five-seater powered by a Turboméca Artouste II turboshaft engine derated to 320 shp. The Alouette II proved an internationally best-selling helicopter, with 923 examples being built for military and/or civil use in 33 countries. Its maximum speed is 115 mph (185 km/h). From 1967 these helicopters were known as S.E.313Bs. On the last day of January 1961 Sud-Aviation flew the prototype of a version of the Alouette II with a Turboméca Astazou IIA turboshaft engine derated to 360 shp. Known as the S.A.318C Alouette II Astazou, production of this stopped in 1975 with the 1,305th and last Alouette II type.

A generally similar helicopter, originally developed for operation

by the forces of India, became the S.A.315B Lama, which first flew on 17 March 1969. Combining the airframe of the Alouette II with the 870-shp (derated to 550-shp) Turboméca Artouste IIIB turboshaft engine of the Alouette III, a Lama took the world altitude record for helicopters to an amazing 40,820 feet (12,442 metres) in 1972, which remains the record today. The Lama is built in France, Brazil and India, in the latter countries under licence. It is known as the Cheetah by the Indian Army. With the founding of Aérospatiale in 1970, the amalgamation of the former Sud-Aviation, Nord-Aviation and SEREB companies, the Alouettes became known as Aérospatiale types.

As a more powerful helicopter to complement the Alouette II, Sud-Aviation developed the S.A.3160 Alouette III. First flown in February 1959, the helicopter remained in production in this form until 1969. It was followed by the similar but strengthened S.A.316B and then the seven-seat S.A.319B Alouette III Astazou (with an Astazou XIV engine, derated to 600-shp). Also built under licence in India, Romania and Switzerland, the Alouette III is today out of production everywhere except Romania.

During the Algerian wars of the 1950s, France experimentally fitted a variety of weapons to its helicopters in an attempt to suppress ground fire. Alouette IIs were particularly useful for this purpose and eventually fired wire-guided missiles. In this respect the French led the world towards the helicopter gunship, but more of this later.

Up to the latter 1950s French helicopters, on the whole, had been small in size. This changed on 19 June 1959 when Sud-Aviation flew the prototype of a large helicopter known as the S.E.3200 Frelon, powered by no fewer than three 800-shp Turboméca Turmo IIIB turboshaft engines. The military version was suitable for troop (24) or stretcher (15) carrying, air-sea rescue, anti-submarine and mine countermeasures, and attack with SS.11 wire-guided missiles. In commercial form it could accommodate up to 24 passengers. However, it did not enter production but instead was developed into the S.A.321 Super Frelon that first flew on 7 December 1962. The Super Frelon is the largest helicopter yet

Above: Following trials with an experimental gunship version of the Alouette III, which carried missiles and a nose gun, helicopters of this type were assigned four AS.11 or two AS.12 missiles. Others carried torpedoes for an ASW role.

Opposite: A ski-equipped S.A. 316B Alouette III of Securite Civile makes a precarious touch landing during training for operation in mountain areas.

Right: Aérospatiale S.A.321G Super Frelon anti-submarine helicopter, also used to patrol the Île Longue naval submarine base while Redoubtable *nuclear submarines are in the area.*

With the Super Puma and Sea King, the Super Frelon is a helicopter equipped to carry and launch the deadly Exocet anti-shipping missile.

built in France and it set a world helicopter speed record of 217.77 mph (350.47 km/h) over a closed circuit that remained unbeaten until 1970. It is currently serving in military and commercial forms. As an airliner it can accommodate from 34 to 37 passengers. But, the Super Frelon is best known for its role in support of the French Navy's *Redoubtable* class of nuclear submarines while at their base on the Île Longue. Production ended in the early 1980s after 99 had been completed.

The word spreads

Most historically significant helicopters produced in the first 15 years after the Second World War came from the U.S.A., Britain, France and the U.S.S.R. Some helicopters produced in other countries are nevertheless, worthy of mention.

Two young engineers, Bernard Sznycer and Selma Gottlieb, both from New York, designed and constructed a helicopter almost immediately after the war as the SG VI-C. From this they developed the intended production prototype SG VI-D, a three-seater powered by a 165-hp Franklin engine and of conventional single main rotor/tail-rotor layout. With the support of financial interests in Canada (collectively known as the Intercity Airlines Company of Montreal). these were probably the first Canadian helicopters.

More important experiments were undertaken in the Netherlands during the early 1950s, where Meijer Drees and Gerard Verhage worked on the development of a small ramjet-powered helicopter at the SOBEH experimental helicopter foundation. The two-seat prototype H-2 was flown for the first time in May 1955 and, in October that year, NHI (Nederlandse Helicopter Industrie) was formed to take over the experimental work of SOBEH. A year after the H-2 flew, the developed H-3 Kolibrie appeared. In 1957 production of the Kolibrie started, with NHI's founding companies, Aviolanda and Kromhout, producing airframes, rotor heads and the TJ-5 ramjet motors of 44 lb (20 kg) thrust.

Up to 20 Kolibries might have been built; after the tenth model they were built with more powerful TJ-5A ramjets. The Kolibrie was simple in design but incorporated one or two notable features. The two blades of the main rotor were designed so that the centre of gravity was forward of the centre

Important experiments conducted in the Netherlands in the 1950s produced the NHI Kolibrie with ramjets to drive the rotor.

The Spanish Aerotecnica AC-11 (Matra-Cantinieau MC-101).

of lift, the air loads thus reducing the angle of attack of the automatic variable-pitch rotor blades and thereby neutralizing vibration. The blades themselves were connected to the rotor head by leaf springs which took the place of the more complicated conventional torsion bearings and flapping hinges. The rather heavy fuel consumption of the ramjet motors was in some part compensated by the use of household-grade paraffin as fuel. Included in the list of optional equipment were two stretchers, crop-spraying and seeding equipment and an external cargo hook to allow loads of up to 827 lb (375 kg) to be lifted. Capable of 70 mph (113 km/h), the Kolibrie proved best suited to agricultural operations.

In Spain in the mid-1950s, Aerotecnica of Madrid developed light helicopters from the French Matra-Cantinieau MC-101 of 1952, which it had purchased at the same time as acquiring the services of Jean Cantinieau.

The first helicopter to be designed, constructed and flown in Brazil was the Beija-Flôr, which made its maiden flight in 1959. Of conventional layout except for the installation of the 225-hp Continental E225 piston engine in the nose and the use of intermeshing twin tail rotors that provided pitch and yaw control and inclined the main rotor plane for forward horizontal flight, it was designed by a team of German engineers at the Centro Tecnico de Aeronautica (CTA) under the leadership of no lesser person than Professor Heinrich Focke.

In the latter 1950s the two-seat HC-2 Heli-Baby light general-purpose helicopter was developed in Czechoslovakia. It was initially fitted with an 83-hp Praga DH engine and was intended for production in slightly more powerful form. In Italy Aer Lualdi produced a neat little Lycoming-engined helicopter as the L.55, to be followed by the more powerful four-seat L.59. The first helicopter of Polish origin was the BZ-4 Zuk, a four-seater powered by a 320-hp Narkiewicz WN-4 radial engine. It was first displayed in August 1956 at the Polish Aviation Day Exhibition.

The first new German helicopter flown after the Second World War was the three-seat Borgward Kolibri I, another design by Heinrich Focke. Piloted by the famous prewar German helicopter pilot, Ewald Rohlfs, it made its first flight on 8 July 1958 on the power of a 260-hp Lycoming VO-435-AlB engine.

The successful German MBB BO 105 helicopter, that remains in production today, was designed and built on the experience gained from the much smaller Bölkow BO 103, an extremely clean-looking single-seater, intended for training purposes in conjunction with the non-flying BO 102 Heli-Trainer. Powered by a 40-50-hp ILO three-cylinder engine, the BO 103 flew for the first time on 9 September 1961.

The Soviet Union becomes a major producer

An important name in Soviet helicopter development was that of Alexander G. Ivchenko. An engine designer, his 55-hp AI-4G engine of

The first post-Second World War German helicopter was the Borgward Kolibri I, designed by Heinrich Focke and piloted by Ewald Rohlfs.

the latter 1940s was his first to be fitted to a helicopter, the Kamov Ka-10, but more of this helicopter later. Of more importance to the overall development of Soviet helicopters was Ivchenko's AI-26, a seven-cylinder radial piston engine that had versions specifically designed for use in helicopters. The very first Soviet helicopter with this type of engine was the G-4, a refined version in Ivan Bratukhin's series of twin-rotor helicopters that had begun with the Omega of 1941. The G-4 was very similar to the earlier Bratukhin helicopters and so carried two such engines. A small batch of G-4s followed the prototypes and early examples were joined by a single B-5 in 1947. The B-5 was ostensibly for the use of the designer and differed from all earlier Bratukhins. It had wings to carry the engines and rotors and a redesigned fuselage to accommodate six passengers in a cabin to the rear of the nose flight deck.

Also in 1947 the Bratukhin B-10 appeared, intended undoubtedly for limited service but remaining a prototype. This helicopter was based upon the same layout as the B-5, but incorporated differences to the airframe and the engines were slightly more powerful examples of the AI-26. Tests may also have been conducted with the B-10 in armed form, although fitted guns would almost certainly have been intended to demonstrate defensive capability while working as an observation, transport or evacuation helicopter. The final helicopters of the series were the two B-11 prototypes of 1948 which had Fa 223-type rounded glazed noses and T-tails married to B-10 type wings and engines. Evaluated against helicopters from Mil and Yak for a communications role, the B-11 was the final Bratukhin helicopter built.

The year 1947 marked the turning point in Soviet helicopter design. In that year Nikolai Ilyich Kamov produced his first helicopter, the Ka-8, which is best remembered as a one-man 'flying motorcycle'. Very few were built but the success of these encouraged further work. Like the Ka-8, the follow-up but more refined Ka-10 of 1949 used twin co-axial rotors, the rotor arrangement that became the hallmark of Kamov helicopters. As noted earlier, the Ka-10 was the first helicopter to use the AI-4G engine and a small number were built for Navy evaluation. However, it took a while for the developed Ka-10M to appear, then supporting twin fins and rudders. A small number were flown on pontoons for sea duties from whalers, coastal patrol and ice-breaking vessels. NATO assigned the reporting name *Hat* to this helicopter.

After the Ka-10 appeared, work began on a much improved helicopter which would offer enclosed accommodation for two persons. A 255-hp AI-14V powered the co-axial rotors and the new fuselage supported a tailplane with endplate fins and rudders. The first Ka-15 flew in 1952 but production did not begin until 1955. This was the first Kamov helicopter to go into large-scale production and service, with hundreds built for military and civil operations. Because of its small size, its military uses were limited to utility, communications and training roles, although on board ships it performed spotting duties and may have had limited experience in an anti-submarine role. Civil versions, used by Aeroflot and exported, included the Ka-15M with two externally mounted stretcher panniers. The maximum speed of the helicopter was 93 mph (150 km/h). NATO assigned the name *Hen* to the Ka-15 and *Hog* to the follow-up Ka-18, which was also built in substantial numbers.

First flown in 1957, the Ka-18 was a four-seat stretched development of the Ka-15 helicopter and had a 280-hp engine in production form. Apart from military versions, the Ka-18 fulfilled many roles including those of passenger, mail and freight transport, agricultural and survey duties. As with the Ka-15, the Ka-18 was probably used on the State Collective Farms.

The currently operated Ka-25 anti-submarine and general-purpose helicopter and its Ka-25K civil flying-crane variant were developed from prototypes that have been spuriously reported as Ka-20s. One was seen carrying dummy air-to-surface missiles at the 1961 Tushino Soviet Aviation Day display. NATO allocated the reporting name *Harp* to the prototypes, while Ka-25 helicopters that went into service during the late 1960s are known to NATO by the name *Hormone*.

Returning to 1947, during that year the bureau headed by Mikhail Mil began work on a single-rotor helicopter that first flew in September of the following year as the prototype Mi-1. Conventional in layout and using rotor blades derived from those of the A-15 autogyro, it became the first Soviet helicopter to enter large-scale series production, in 1950, and was rushed into military service from 1951. Like the contemporary Yakovlev Yak-100 (which to all intents and purposes looked like a straight Soviet copy of the Sikorsky S-51 but was, in fact, an entirely separate development with an M-11 engine of 140 hp), it had a bulbous cabin with the rotor pylon-mounted above and a long boom supporting a tail rotor. However, whereas the Yak-100 was a three-seater, the Mi-1 could accommodate up to four persons and indeed beat the Yak to production orders.

Many variants of the Mi-1 were produced for military and civil use, fulfilling most of the roles then recognized as being suitable for small helicopters. It was widely exported and in 1955 production was transferred from the Soviet Union to Poland where WSK-Swidnik introduced the SM-1 variants. The standard engine was a 575-hp Ivchenko AI-26V, which allowed a maximum speed of 118 mph (190 km/h). NATO referred to the Mi-1 as *Hare*.

The Mi-2 was *not* the follow-up helicopter to the Mi-1, as might be expected, but a modernized deriva-

Above: A Soviet Navy Kamov Ka-25 anti-submarine helicopter photographed while flying over the Sea of Japan in 1983. Search radar is housed inside the undernose radome, and a towed MAD (magnetic anomaly detector) is carried and dipping sonar is accommodated to the rear of the main cabin.

Left: The first Kamov helicopters to enter large-scale production were the Ka-15 and Ka-18. Here an Aeroflot Ka-18 poses in front of a military Ka-15.

Right: The Mil Mi-1 was built in Poland as the SM-1, the many variants including the SM-1WS ambulance with two external stretcher panniers.

Left: This poor quality photograph is nevertheless interesting as it depicts a Soviet Mi-2 operating in support of main battle tanks on manoeuvres.

Right: Mil Mi-4 operating in the difficult conditions of India.

Below: The larger twin-turbine Mi-2 followed the SM-1 into production in Poland.

tive, conceived and developed in the Soviet Union but handed over to WSK-Swidnik for final development and then full production. The first Mi-2 flew in September 1961. It was an enlarged helicopter which could accommodate up to eight passengers in addition to the pilot. To cope with the increase in weight, the piston engine was superseded by two locally built Isotov GTD-350 engines of 430 shp each. Production in Poland began in 1965 and continues today (in updated form). It has been estimated that about 4,000 Mi-2s have been built both for home use and for export, with the majority going to the Soviet Union. Both civil and military versions have been produced. Some Polish Air Force helicopters carry *Sagger* anti-tank missiles on short fuselage-mounted pylons. Civil Mi-2s include specialized versions for firefighting, ambulance and agricultural work. NATO uses the name *Hoplite* for the Mi-2.

Between 1952 and 1969 it is believed that Mil produced about 3,500 helicopters of the Mi-4 type, the true successor of the Mi-1. Development of the Mi-4 was to the express order of Stalin, who let it be known that he expected a dramatically improved helicopter for flight testing within one year of September 1951. Mil's task was to produce a helicopter capable of carrying a 2,646-lb (1,200-kg) payload. Yakovlev was, at the same time, instructed to produce a helicopter with a 5,292-lb (2,400-kg) payload, which resulted in the Yak-24.

Though difficult, Mil's task was somewhat easier than Yakovlev's and it was amply fulfilled with the Mi-4. Whether or not the design of this Soviet helicopter was influenced by the American Sikorsky S-55 is not of very great importance, as the Mi-4 was built from Soviet-developed components. Larger than the S-55, the Mi-4 adopted a scaled-up Mi-1 rotor driven by a 1,700-hp Shvetsov ASh-82V radial piston engine. Exclusive to the Soviet helicopter were clamshell loading doors to allow straight-through loading of bulky freight or of a vehicle into the main cabin and an undernose gondola on military versions for a navigator/observer. The Mi-4 proved capable of lifting a 3,836-lb (1,740-kg) payload, although 3,525 lb (1,600 kg) was normal.

The first Mi-4 prototype to fly became airborne in May 1952 and the initial military production examples entered service the following year. Apart from anti-submarine and general-purpose duties (14 troops, 8 stretchers and so on), the military Mi-4 could be armed for Army support with a machine-gun in the gondola and rockets. Among the civil versions were those for carrying 10 passengers, firefighting and agricultural work.

The first helicopter built in China was the Mil Mi-4, assembled at the Harbin works and known in that country as the Z-5. Both Soviet and Chinese versions were allocated the reporting name *Hound* by NATO. The Z-5 remained in production in China for two decades from 1959 and about 1,000 were built for military and civil operation. All the Z-5s used a locally built version of the Shvetsov engine and so shared a maximum speed of 130 mph (210 km/h). Since 1979, Harbin has been converting some

Mi-4/Z-5s to use the Pratt & Whitney Aircraft of Canada PT6T-6 Turbo Twin Pac turboshaft engine.

In the 1980s the Mi-4 is still flown in many countries but later and better helicopters have long since replaced it for all the more important roles in the Soviet Union, where, in military form, it is used mainly as an electronic countermeasures helicopter to jam communications.

Very different from the Mi-4 is the Mi-6, which first appeared in September 1957 to meet military requirements for a VTOL aircraft capable of carrying bulk cargo in strategic operations. Also, the airline Aeroflot required a civil helicopter to transport bulk cargo to developing regions of the Soviet Union. This capability was amply demonstrated when the helicopter set no fewer than 14 world records in its class, including one that saw the Mi-6 lift a 44,350-lb (20,117-kg) payload, a load slightly greater than the normal maximum payload of an An-12 transport aeroplane. Even today the Mi-6 retains world records, although these are mostly connected with speed. Indeed, the Mi-6 is capable of 186 mph (300 km/h) on the power from its two 5,500-shp Soloviev D-25V turboshaft engines which drive the single five-blade main rotor. Under normal operations the Mi-6 can carry internal and external payloads of 26,455 lb (12,000 kg) and 17,637 lb (8,000 kg) respectively. It is thought that about 800 helicopters of this type have been built for military (70 troops, 41 stretchers or other loads) and civil (65–90 passengers) uses in the Soviet Union and for export. It is thought that Soviet tactical forces operate approximately half this number. From the Mi-6 was developed the Mi-10 and the more specialized Mi-10K heavy flying-crane helicopters that were first seen in public in 1961 and 1966 respectively. These later helicopters do not use the Mi-6's detachable wings, which provide about one-fifth of the Mi-6's 'lift' in normal cruising flight.

The first helicopter bearing the Yakovlev name was the EB, a small two-seat co-axial machine that flew in 1947. Later, having had the Yak-100 rejected for production in favour of the Mil Mi-1, Yakovlev took up the difficult challenge of producing a helicopter, capable of lifting a payload of 5,292-lb (2,400 kg), twice that expected of the Mi-4, and to have it flying within a year of September 1951. To speed development, Yakovlev sought and was given approval to use two of the Mi-4's engine/rotor combinations which were married to a Piasecki-type long fuselage. The forward and rear portions of the square-section fuselage were skinned with metal; the centre portion was fabric covered. Fabric also covered the large fin on which the rear four-blade rotor was mounted. Heavily upswept tailplane surfaces were incorporated in the design. When production eventually got underway, later series-built examples were skinned entirely with metal and the dihedral tailplane gave way to a new braced tailplane with greatly reduced dihedral and endplate fins.

The Yak-24, as it became known, was actually designed by N. K. Skrzhinskii with the assistance of a workforce of 10. Tethered flight tests began with a prototype in July 1952 and, after serious vibration had been cured, the helicopter entered production in Leningrad.

The Yak-24 was first seen in public at the August 1955 Soviet Aviation Day display at Tushino. On 17 December a Yak-24 set two world records for its class of helicopter; the first saw a 4,410-lb (2,000-kg) load lifted to a height of 16,673 feet (5,082 metres) and the second a 8,820-lb (4,000-kg) load taken to 9,521 feet (2,902 metres). The later appearance of the Mi-6 saw the latter record beaten before the end of the decade.

Outside the Soviet Union, it was thought, in 1959, that the Yak-24 was about to enter Aeroflot service as a 30-passenger transport. This, in fact, did not happen. The civil model would have been the Yak-24A. A VIP variant was also to remain a prototype as the Yak-24K. Therefore, all of the 100 or thereabouts Yak-24s built went into military service. It is believed that the Yak-24 (NATO *Horse*) was capable of carrying a military load of up to 8,820 lb (4,000 kg), which could include two GAZ-69 jeeps, one GAZ-69 vehicle towing a twin-barrel 14.5-mm anti-aircraft gun and its crew of six, three medium trucks, freight or up to 40 troops.

Stalin's insistence on improving the VTOL capability of the Soviet forces in as little time as possible and without concern over the expense had paid off handsomely with the Mi-4 and Yak-24, though Yakovlev thereafter concentrated on fixed-wing combat aircraft and only much later VTOL combat aeroplanes and civil airliners. Mil, on the other hand, never looked back and is today one of the two largest producers of helicopters, matched only by Bell Helicopter Textron in the U.S.A.

Above: The last helicopter by Yakovlev was the Yak-24, an ambitious twin-engined/rotor transport designed to Stalin's order.

Top: An early example of the long-legged Mil Mi-10 flying-crane helicopter, which uses closed circuit TV to scan the payload during operations.

Opposite top: Until the late 1960s the Mil Mi-6 heavy transport helicopter was the largest helicopter in the world. Here, one lifts a Soviet space vehicle.

Korean Angel to Vietnam Veteran

Once helicopters with some practical value became available, their first and most important customers were the military. The earliest series-built helicopters took on varied non-combat roles, with successive models from any particular manufacturer usually offering enhanced load-carrying capability. By the close of the 1940s Sikorsky had taken load-carrying capacity to a new high with its S-55, while in the Soviet Union the Mi-1 was in the early stages of mass production. On the other end of the size scale, the Bell Model 47 was finding expanding markets in military and civil fields, and was the first fully practical helicopter for agricultural work.

Two Sikorsky H-5s of the 3rd Air Rescue Squadron, USAF, rest between missions in Korea in February 1951.

Then, on 25 June 1950, North Korean forces began a dawn crossing of the 38th Parallel into South Korea. During the two days of 27 and 28 June the United Nations Security Council debated the Korean crisis and passed a resolution calling upon its members to assist the besieged South in any way possible. On 3 July U.S. jet fighters became engaged in combat for the first time when Grumman Panthers flew from the Navy aircraft carrier USS *Valley Force* against North Korean forces. By 27 July a huge expansion programme for the U.S.A.F. had been announced (which naturally encompassed increased production of helicopters), followed by announcements to expand French forces on 6 August and U.K. forces on 30 August. In November U.S. jet fighters met Chinese-flown Soviet Mikoyan-Gurevich MiG-15 jet fighters in combat and thereafter Chinese and North Korean MiGs and other fighters were regularly fought in air-to-air combat. Other United Nations' warplanes used to assist South Korea were Gloster Meteors flown by the Royal Australian Air Force. On 16 December President Truman proclaimed a state of emergency in the U.S.A. to cope with the raging war and announced that the U.S. forces would be doubled. On 1 January 1951, reinforced by 400,000 Chinese troops, North Korea began a new advance on the South.

It was against this background that several countries sent forces to fight with South Korea under the banner of the United Nations. This war was not only the biggest since the Second World War but the first and only really major conflict to

involve United Nations forces in the actual fighting. Apart from being the backdrop for the first air-to-air combats between jet aircraft and for the first-ever use of the terrifying napalm bomb, the Korean War helped to establish the helicopter as a vital workhorse not only just behind the front line but even over enemy-held territory. Although not initially used for actual combat, American helicopters of all production types in military use carried troops, supplies, guns and ammunition to remote and otherwise inaccessible positions, rescued 'downed' pilots (often from behind the enemy's front line under the protective firepower of fighter aircraft), and performed other varied duties. However, perhaps the helicopters greatest feat of all was to air-lift more than 23,000 United Nations casualties to field hospitals over the three-year war, thus making possible the lowest percentage of wounded to die in the history of warfare. Of these, no fewer than about 18,000 were carried by Bell H-13s (Model 47s), each helicopter supporting two externally mounted stretchers. For its remarkable work in saving lives, the H-13 earned the nickname *Korean Angel.*

Fifties fighting

While Korea occupied most headlines, British Commonwealth forces were concurrently fighting a jungle war in Malaya against communist guerillas. This war, which outlasted that in Korea by many years, was one of attrition, won mainly by depriving the guerillas of the support of the population. The Dragonfly helped greatly in this task. Used initially as a casualty evacuation aircraft, it also assisted in setting up fortifications in the jungle which were used to afford protection for the surrounding populations.

From the 9th Regimental Combat Team, 2nd US Infantry Division clearing station in Korea, wounded are evacuated by Bell H-13 'Korean Angel'.

Right: Westland Whirlwind 2s over Kuala Lumpur, Malaya.

Below: A Belvedere helping to construct a pontoon bridge.

Dragonflies were joined, from 1954, by Bristol Sycamores, which often carried troops to mountainous positions, a role which was even better-suited to the Westland Whirlwind because of its greater capacity. The Whirlwind is also remembered for its many rescue missions in Malaya.

British helicopters were flown in other trouble spots around the world, including Cyprus and Tanganyika. In Tanganyika the larger Westland Belvederes transported commandoes from HMS *Centaur* during operations in 1963. Belvederes were again in action in Brunei in the 1960s.

Meanwhile the French had fought hard and long in Algeria during the 1950s and it was here that the helicopter became a gunship. As early as 1942 a prototype of the Sikorsky R-4 had been used at Wright Field for armed trials. This first-ever armed helicopter initially hovered over a target while the passenger released, by hand, a small 25-lb practice bomb. This trial was followed by others with a 20-mm cannon. But armament was not adopted for this type of helicopter. Indeed, even during the Korean War progress towards arming helicopters was stunted. But the French Army and Air Force, operating their Alouettes and S-55s in Algeria under increasing enemy ground fire, quickly learned the value of machine-guns to offer defence

Piasecki H-21s served with distinction in combat with the French Army in Algeria.

during battlefield operations. It was a short step from this to installing air-to-ground rockets and later wire-guided missiles not only to suppress ground fire but attack enemy emplacements and other targets.

Even after the French had proven that the helicopter could be a platform for weapons and was ideally suited to operations in difficult terrain, little progress was made elsewhere along the same line. Back in the mid-1950s Bell had been selected to provide the U.S. Army with a new helicopter for utility, casualty evacuation and training roles and this, the Model 204, was ordered into production as the H-40, but was subsequently redesignated HU-1 and named the Iroquois; deliveries to the U.S. Army began in 1959. Power for each helicopter, capable of accommodating eight troops or 4,000 lb (1,815 kg) of freight, was provided by a Lycoming T53-L-1A turboshaft engine derated to 700 shp. Becoming UH-1As from 1962 under a new inter-service system, these were among the first helicopters to serve in Vietnam.

Above: Bell UH-1B 'Huey', a seven troop helicopter powered by a 960 shp or, later, a 1,100 shp Lycoming T53 turboshaft engine.

Left: A Hiller H-23 with stretcher panniers comes in to land near French forces fighting in Vietnam during the 1950s.

Vietnam – the gunship evolves

The United States of America's involvement in Vietnam began when the decision was taken to offer support to the government of South Vietnam in terms of military advisory forces and assistance following fresh insurgencies into the South by forces of the North. The Americans were not the first to offer assistance, for during the 1950s the French had fought in Vietnam. U.S. assistance included training and support of the South Vietnamese Army, and so, in December 1961, the 57th Transportation Light Helicopter Company arrived in Tan Son Nhut. Flying Piasecki H-21s, the company was used, first, to train Vietnamese troops to embark from helicopters in a drop zone in minimum time, and this led to the first actual support mission on 23 December, when H-21s air-dropped 360 Vietnamese infantry for a search and destroy mission.

Non-combat support from U.S. forces continued until 1965, with Bell UH-1s (known as Hueys) having taken over from the H-21s in 1964. However, well before this, in 1962, 13 UH-1As operated by the Utility Tactical Transport Helicopter Company in Vietnam had been armed with two 0.30-inch machine-guns and 16 2.75-inch rockets, each as an experimental armed escort helicopter. Sixteen UH-1Bs, a follow-up version of the Iroquois 'Huey', delivered to the Army from March 1961, were the next to be armed experimentally;

Above: The international military export version of the Chinook is the Model 414. Three were acquired for the 5th Helicopter Transport Battalion of the Spanish Army, operating from Colmenar, Viejo, during 1982.

Right: A Chinook of the 1st Squadron, 9th Cavalry, 1st Cavalry Division (Air Mobile) dumps 55-gallon drums of CS riot-control gas into suspected Viet-Cong emplacements in July 1967.

Left: The water-tight lower fuselage of the Chinook allows water landing capability, demonstrated by this CH-47C. Note the two 3,750 shp Avco Lycoming T55-L-11C turboshaft engines carried in pods on the sides of the rear rotor pylon.

this time each was armed with six French Nord SS.11 wire-guided anti-tank missiles. By 1965 General Electric had developed a 40-mm turret-mounted grenade launcher for helicopters.

In 1965 U.S. forces were fully committed to supporting the government in the South and now joined in the actual fighting. The UH-1B and later versions of the Huey were vital to the war effort and many were armed as escorts both for troop- and supply-carrying helicopters, as well as for rescue helicopters. It was the task of these armed helicopters to attempt to defend against and then suppress enemy ground fire during these missions and offer fire support against armoured attack. Armament varied but two electrically controlled guns firing from each door opening and 48 rockets became typical for each Huey.

One of the most sophisticated versions of the Model 204 was the UH-1M, used by hunter-killer helicopter groups in 1970 to detect and attack targets at night using the Hughes Aircraft INFANT night fighter and night tracker system.

Early in the Iroquois production programme, a stretched version appeared as the Model 205. It used a more powerful version of the T53 engine and accommodated up to 14 troops. This, too, was widely used in Vietnam and, like the Model 204, served with the other U.S. armed services. It was also exported and built in commercial form.

During the Vietnam war the helicopter truly became a gunship,

Above: Machine-gun armed UH-1D and UH-1H (Model 205 Iroquois) 'Huey' helicopters, used by the US Army on search and destroy missions over Vietnam enemy supply routes.

Top right: Of the many foreign forces to take delivery of Iroquois helicopters can be counted the Venezuelan Air Force. Four of its UH-1Ds are pictured here.

Bottom right: A Brazilian Air Force UH-1H search and rescue helicopter.

FAV
CHRI
FAV

FÔRÇA AÉREA BRASILEIRA
SAR
SAR

Left: A rather 'bent' Douglas AD-5 Skyraider is salvaged by a Sikorsky CH-54A Tarhe of the 1st Cavalry Division (Air Mobile) in Vietnam in December 1965.

Below: South Vietnamese troops wade through the mud to an awaiting Chinook.

so much so that the U.S. Army set up a special unit called ACTIV (Army Concept Team in Vietnam) to evaluate armed operations. Many different types of American helicopters operated in Vietnam. Individual assault companies comprised three transport helicopters and a Huey gunship. Assault support companies used Sikorsky CH-54 Tarhe (S-64) flying-cranes and Boeing Vertol H-47 Chinooks. The Chinook, originally a Vertol design, first flew on 21 September 1961, by which time it had already been ordered by the Army Air Corps as its first standard battlefield mobility helicopter. Boeing Vertol's earlier Model 107 did not enter Army service but was selected by the U.S. Marine Corps and the U.S. Navy as the H-46 Sea Knight. This also became a Vietnam veteran. Chinooks are still produced by Boeing Vertol today and have been successfully exported over the years.

American helicopters operating in Vietnam also included Hiller H-23s, Bell H-13s, Bell OH-58A Kiowa military observation derivatives of the Model 206A JetRanger

Left: Sikorsky S-64 Skycranes have also been purchased for commercial operations. Here a Skycrane transports a massive load of timber.

Right: US Marine Corps CH-46 Sea Knight extracts a reconnaissance team during the Vietnam fighting of 1967.

Below: US Marine Corps Sea Knights modified to use 1,870 shp General Electric T58-GE-16 turboshaft engines, and with greater resistance to combat and crash damage, are known as the CH-46Es. 273 Sea Knights are being updated in this way.

civil light helicopter, and Hughes OH-6A Cayuse observation helicopters (from which civil Model 500 series helicopters were derived). But initially the United States was not equipped for this type of war and found on many occasions that it had to rethink its strategy. For example, old and slow propeller fighter-bombers were often of more value for striking at the illusive enemy and low-priority targets than multi-million-dollar jets, especially during operations that put a modern jet of Phantom II type and its crew at risk.

With its own armour, heavy guns and modern jet fighters for support, the Viet-Cong on the ground could live in the jungle away from bases for long periods, eating food easily carried or supplied by bicycle and penetrating deep into the South using all the natural cover available. It was under these conditions that the U.S. forces attempted to contain the war, striking at military targets and communications in the North and South using bombers and fighter-bombers but never attempting to invade the North, as such an act might have had the result of bringing the Chinese into the actual fighting as had happened during the Korean War.

One of the most important developments in aviation to result from American involvement in Vietnam was the deployment of the world's first specialized and purpose-designed armed close-support and attack helicopter. From the outset

Above: Boeing Vertol CH-113 Labradors, similar to CH-46A Sea Knights, were delivered to the Canadian Armed Forces for search and rescue missions.

Above: The Bell OH-13S was one version of the H-13 flown in Vietnam, used as an observation helicopter and powered by a 260 hp Lycoming TVI-435-25 piston engine.

Left: The OH-58 Kiowa was developed as a military light observation and training helicopter from the Bell Model 206A JetRanger.

Right: Delivered in 1976, the Austrian Army received twelve OH-58B Kiowas.

of U.S. involvement in late 1961 it became clear that helicopters were susceptible to ground fire, despite German and American mock combats during the Second World War which had shown manoeuvring rotorcraft to be difficult targets to then-modern fighters. Not altogether surprisingly Bell Helicopter was the first to recognize fully this problem and in 1962 used its Iroquois as the basis for a new specialized armed helicopter which it named the Iroquois Warrior. In place of the standard UH-1 fuselage, the Warrior had a new slim body with tandem seating for a crew of two, the rear seat raised above that of the forward position to offer excellent vision. The small frontal area of the Iroquois Warrior meant that it would offer a smaller target to enemy ground fire, while manoeuvrability and aerodynamic efficiency would be greatly enhanced. The lighter body also allowed more armament to be carried, a particularly important feature as the armed UH-1As used in trial operations were having difficulty keeping up with the unarmed Hueys.

A mock-up of the Iroquois Warrior was completed in June 1962 and, with only a couple of months to go before the Army expected its initial deliveries of the new Chinook medium transport helicopter, interest was shown in the helicopter as an escort. However, while appreciating the concept, the Army considered that an existing helicopter could be adequately modified into a special-

Right: Hughes OH-6A Cayuse observation helicopter in the service of the US Army.

The all-important cockpit section of Bell's revolutionary Iroquois Warrior.

12931
UNITED STATES ARMY

Right: The first-ever purpose-designed attack helicopter was the experimental Bell Model 207 Sioux Scout.

Below: Kaman HH-2C Seasprite, considered as an interim AAFSS helicopter, carrying a Minigun chin turret and with the waist gun clearly seen.

Armed version of the Boeing Vertol CH-47A Chinook, four of which were evaluated in Vietnam as ACH-47As in 1966. Carrying a new type of steel armour for protection, each ACH-47A could be armed with a 40 mm M-5 automatic grenade launcher on the nose, pylon-mounted guns and rockets, plus four flank machine-guns.

ized armed machine and in the 1964 fiscal year millions of dollars were allocated for development along this line. One helicopter that emerged was a version of Kaman's UH-2 Seasprite, now sporting both an armament turret at the nose and a pair of stub wings to which further weapons could be attached.

When the Secretary of the Army decided to increase the required maximum speed of an attack helicopter by 49 knots to a hitherto unimagined 200 knots (230 mph/370 km/h), Bell became more-than-ever convinced that only a new helicopter could meet the requirements. Already Bell had begun work on a flying demonstrator of its tandem-seat armed helicopter but, as this had to be company funded and produced in haste, it was based on the H-13. As the Model 207 Sioux Scout, the first-ever purpose-designed attack helicopter demonstrator took to the air in July 1963.

If anything, Bell had been too convincing in its arguments for a specialized attack helicopter to join U.S. Army units. In 1964, the Army decided that it required a highly armed and fast helicopter and issued a specification for an AAFSS (advanced aerial fire support system) helicopter, the specifications for which went beyond Bell's own ideas. However, Bell had been keeping a close eye on world events and the AAFSS programme, and realized that a specialized attack helicopter would undoubtedly be needed for urgent service before AAFSS types became available. In 1965, therefore, the company began work on an interim type, smaller than the Iroquois Warrior but still using components of the UH-1C. This, too, was company funded. In

August 1965 the U.S. Army set up a committee to find an interim attack helicopter while the full AAFSS programme progressed. Because of its earlier commitment, Bell was able to fly the prototype of its new Model 209 HueyCobra on 7 September. In November that year the HueyCobra was evaluated against armed versions of the Sikorsky S-61A and Kaman UH-2 and on 13 April 1966 110 examples of the HueyCobra were ordered from Bell, the first order for a specialized attack helicopter.

Production AH-1G HueyCobras were delivered to the U.S. Army from June 1967. Operational deployment began in August and altogether 1,078 AH-1Gs were built over several years. Possessing a fuselage only 3 feet 2 inches (0.97 metres) wide and a maximum speed of 219 mph (352 km/h), the HueyCobra was indeed a difficult aircraft to hit from the ground. Its VTOL operations, combined with firepower from an Emerson Electric TAT-102A nose turret housing a six-barrel 7.62-mm Minigun and underwing weapons, meant that the U.S. Army could make surprize attacks on ground targets with real muscle; escorting unarmed helicopters was also well within its capabilities. Power for the AH-1G was provided by a 1,400-shp Lycoming T53-L-13 turboshaft engine derated to 1,100 shp, as fitted to later versions of the Iroquois.

As the war went on armament fitted to the AH-1G increased, an XM-28 turret with two Miniguns or two 40-mm grenade launchers or one of each replacing the original turret. Meanwhile, in November 1968 production began of the Hughes Aircraft TOW heavy anti-tank missile which could be fired from Army armoured vehicles and

To the Bell Model 209 AH-1G HueyCobra went the distinction of becoming the first specialized attack helicopter to enter production and service. By August 1967 AH-1Gs were being deployed operationally in Vietnam.

Above: HueyCobras during their first month of operation in Vietnam, parked in their burms at Bien Hoa as protection against possible mortar fire.

Left: As part of the early flight training of HueyCobra crews in Vietnam in 1967, combat missions were flown. This AH-1G fires rockets and its TAT-102A Minigun at a target 43 miles (69 km) from Saigon.

jeeps. In 1972 tests were conducted to evaluate its usefulness as a weapon for the HueyCobra and eight AH-1Gs were modified under ICAP (Improved Cobra Armament Program). Helmet sights were provided by UNIVAC.

To help meet an expected offensive by the Viet-Cong, some of the eight were sent to Vietnam in late April 1972, after the crews had fired only one TOW missile each in training. They were sent to Kontum, where by 27 June they had flown 77 missions and hit 62 point targets, destroying 39 armoured vehicles, howitzers and trucks without loss to themselves. This success led to the decision to convert more AH-1Gs to anti-armour AH-1Qs, leading, in turn, to further modifications and new construction to AH-1S-series type. Meanwhile, a twin-engined version had been developed for the U.S. Marine Corps as the AH-1J Sea Cobra, delivered from 1970, each with a Pratt & Whitney Aircraft of Canada T400-CP-400 coupled free-turbine turboshaft engine derated to 1,250 shp. These were followed by improved AH-1Ts. AH-1Gs and AH-1Js have also been exported to a small number of countries, while perhaps the HueyCobra's most unusual task today is performed by two ex-U.S. Army AH-1Gs that are used by the U.S. Customs Service for night interception of criminals.

Above: Following the delivery of two US-built AH-1S HueyCobras to the JASDF for evaluation, Fuji of Japan is undertaking production of the helicopter under licence from Bell. Here a TOW missile blasts from its launcher.

Right: AH-1S HueyCobra carrying TOW missiles.

Above: An improved version of the AH-1J for the USMC became the AH-1T Improved SeaCobra, using the 2,050 shp Pratt & Whitney T400-WV-402 power plant, Model 214 rotor system, and some KingCobra technology in a lengthened fuselage. These changes allow increased weapon carrying capability. The first AH-1T was delivered in 1976. This, the 45th example built, has been re-engined with two General Electric T700-GE-700s for research into future Cobra models and is known as the AH-1T+ SuperCobra.

Left: No fewer than 202 Bell AH-1J SeaCobras were delivered to Iran from 1974, before the revolution, this one standing alongside another Bell helicopter acquired by Iran, a 15-troop Model 214A.

AAFSS

Meanwhile, the actual contract for a full AAFSS helicopter had gone to Lockheed in 1965 and on 21 September 1967 the first of 10 prototype AH-56A Cheyenne helicopters was flown for the first time. On 7 January 1968 the Department of Defense approved production of 375 but in 1969 the complete programme was cancelled before any had reached service. Powered by a 3,435-shp General Electric T64-GE-16 engine, the Cheyenne would have been capable of 253 mph (407 km/h) and featured a pusher propeller at the tail in addition to the normal main rotor and anti-torque rotor. Why had the Cheyenne been cancelled after the U.S. Army had placed such emphasis on the AAFSS programme? Simply because the HueyCobra was tackling a difficult task in Vietnam more than adequately and the Cheyenne appeared too expensive to put into mass production given its advanced technology.

Bell once again took the initiative in 1971 and offered the Army its Model 309 KingCobra as a HueyCobra replacement. The prototype, built from company funds, made its maiden flight on 19 September that year. However, it was not taken up by the U.S. Army and Sikorsky's S-67 Blackhawk (which took the world speed record for helicopters to 220.888 mph/

Above: Lockheed AH-56A Cheyenne prototype, built to the US Army's AAFSS requirements.

Left: Bell Model 309 KingCobra.

Right: In late 1962, Lockheed flew its XH-51A high-performance helicopter with a 500 shp Pratt & Whitney Aircraft of Canada T47 turboshaft engine. Built to US Army/Navy contract, in Compound form with an auxiliary J60-P-2 turbojet engine on the fuselage side (as seen) and short wings, it flew at 302.6 mph (487 km/h) in 1967, thus becoming the first helicopter to exceed 300 mph. However, the XH-51A Compound was not eligible for a world record in a helicopter category.

ARMY

1263

Right: Sikorsky S-67 Blackhawk, once the fastest helicopter in the world.

Below: Bell YAH-63, a contender for AAH orders.

Left: An Apache blasts a Hellfire test round at a simulated target.

Above: The winner of the US Army's AAH programme was the Hughes YAH-64 Apache which, from 1984, was being delivered in production form as the AH-64A.

355.485 km/h) was also rejected for production. Nonetheless, a more advanced helicopter to replace the HueyCobra was not overruled and a new specification was drawn up for the U.S. aircraft industry under the AAH (Advanced Attack Helicopter) programme. The specification required that the new helicopter be capable of day, night and bad weather operations against armour and of operating away from base for extended periods while in the front line of battle. To this end Bell developed its Model 409, U.S. Army designated YAH-63, and Hughes developed its Model 77 Apache, Army designated YAH-64. In 1976 it was announced that the Hughes helicopter had been selected for production and in 1984 the first production AH-64A Apaches were delivered to the U.S. Army. The first operational unit will be the 6th Cavalry Combat Brigade at Fort Hood, Texas.

Powered by a 1,696-shp General Electric T700-GE-701 turboshaft engine, the tandem two-seat Apache is capable of 186 mph (300 km/h). Under its fuselage is carried a Hughes Helicopters M230A1 Chain Gun 30-mm cannon and stub-wing armament can include up to 16 Rockwell Hellfire anti-tank missiles or 76 2.75-inch rockets.

Overleaf: An impressive display of the weapons available to the Apache, including 1,200 rounds of 30-mm cannon shells, seventy-six 2.75-in rockets and sixteen Hellfire anti-tank missiles.

Left: Hughes Model 500MD Defender carrying a pod with two TOW missiles and an 0.5-in machine-gun, one armament arrangement available to this specialized attack version of the Model 500 for expanded capabilities.

Below: Thirty Defenders were acquired by Israel, while exports to other nations have included anti-submarine and anti-shipping versions.

Outside the U.S.A.

The use of armed helicopters for anti-armour and other missions spread quickly to other countries, most of which now operate versions of existing helicopters armed with guns, rockets or missiles. Some manufacturers, like Hughes Helicopters in the U.S.A, Aérospatiale in France, MBB in Germany, Agusta in Italy and Westland in the U.K., offer specialized anti-tank and other armed versions of existing helicopters. These are popular because they are less expensive to purchase and can be used for other duties if required. However, European manufacturers have also designed specialized tandem two-seat machines, though none are as yet beyond the development stage.

Soviet helicopters are often fitted with some armament as a matter of course. For example, a number of Mi-6s have a 12.7-mm gun at the nose. The Mi-8 transport, that first appeared in public in 1961 and has since gone into service as an assault transport and civil helicopter airliner (up to 24 seats), has also appeared in fully armed form under the NATO reporting names *Hip-E* and, for export, *Hip-F*. The former is recognized as the world's most heavily armed helicopter. Each example in service with Soviet tactical air forces has a gun at the nose, 192 air-to-ground

Right: A version of the Defender with a mast-mounted sight is the Model 500MD/MMS-TOW, here seen with a missile leaving its launch container.

Above: Another version of the BO 105 for military duties is the BO 105M (VBH), a liaison and observation helicopter delivered from 1980 to the Federal German Army.

rockets carried in six packs supported from racks on the cabin sides and four *Swatter* anti-tank missiles.

The first purpose-designed heavily armed Soviet helicopter was the Mil Mi-24, of which production has totalled more than 1,000. This figure includes those exported to Warsaw Pact countries and other nations. It is known to NATO as *Hind*. The first known deployment went to East Germany in 1974. The initial models appeared as armed assault helicopters with a crew of four, room for eight troops in the main cabin, a machine gun in the nose and missiles or rockets under auxiliary wings. From an attack viewpoint, the development of *Hind-D* and then *Hind-E* is particularly important as these later models of the Mi-24 have recontoured noses to accommodate a crew of only two in tandem and are specifically gunship helicopters. The *Hind-D* has an undernose turret, housing a four-barrel gun which can be fired against other helicopters as well as ground targets. Under the wings four missiles and four rocket pods, bombs or other weapons can be carried. *Hind-E* is even more formidable, carrying four new *Spiral* missiles in place of the *Swatters* and has a twin-barrel cannon on the fuselage side instead of a nose turret. The *Hind*, powered by two 2,200-shp Isotov TV3-117

Left: West Germany's MBB BO 105P (PAH-1), an anti-tank version of the BO 105, armed with six Euromissile Hot missiles and with an SFIM stabilized target acquisition and tracking sight on the fuselage above the co-pilot's seat. This type of helicopter serves with the Federal German Army.

Below: A British Army TOW-carrying Westland Lynx AH.Mk 1.

Above: Mil Mi-8 in Finnish service.

Left: Mil Mi-24 Hind-A, *a heavily armed assault helicopter with the original cockpit configuration.*

Right: This head-on view of a Mi-24 Hind-D *gives a good indication of the revised tandem crew layout, in addition to showing the chin four-barrel machine-gun turret and underwing pylons for rockets and missiles (the latter not carried on this occasion).*

Below: The outcome of warfare is dramatically illustrated by this photograph, showing three Argentine Army Iroquois helicopters and other military equipment abandoned at Moody Brook near Stanley on the Falkland Islands.

Right: Designs are being produced in the USA to the US Army's future LHX (Light Helicopter Experimental) programme. The helicopters will be required for service from the mid-1990s, one form as a utility helicopter to carry a small number of troops or other loads, the second as a scout and attack helicopter. An artist's impression of a winged Hughes NOTAR-type helicopter is shown here, in scout and attack LHX configuration.

turboshaft engines, is capable of 199 mph (320 km/h), while a record-breaking version known as the A-10 holds the world speed record for helicopters at 228.9 mph (368.4 km/h). Soviet and Afghan *Hinds* have been flown operationally in Afghanistan since December 1979.

It is now known that the Soviet Union has also developed a new attack helicopter of Apache type, known as the Mil Mi-28 (NATO *Havoc*).

Up to the year 2000

Special interest at present surrounds the development of the world's first helicopters intended for actual air-to-air combat. Little has been reported up to the present time regarding progress towards such helicopters, which are not likely to become part of the military aviation scene for some years to come. Work on the design of air-combat helicopters has taken place in the United States but it is believed that the Soviet Union has taken an early lead and might even have prototypes flying. What configuration such helicopters will have can only be surmised, although the accompanying artist's impression from Hughes Helicopters may lead one to believe that they will be of compact form, capable of high speed and excellent manoeuvrability and missile armed.

Above: Contending helicopter designs for the US Army's LHX programme are likely to come from Hughes, Bell, Boeing Vertol and Sikorsky, not least because LHX helicopters could eventually comprise the bulk of the US Army's helicopter force, replacing many conventional types. It is thought likely that if the speed requirement is determined at 200 knots and not 300 knots, Bell and Boeing Vertol could join forces to submit a tilt-rotor helicopter of a type perhaps similar to this artist's impression.

Passengers to Petroleum

What can be termed the modern era for helicopters really began in the late 1950s, when designs were produced that eventually led to some of the military and civil helicopters that are familiar in our skies today.

One of the most versatile designs of the late 1950s was Sikorsky's S-61, developed to U.S. Navy contract for an amphibious anti-submarine helicopter that could perform both 'hunter and killer' roles. At that time, Sikorsky HSS-1s (U.S. Navy S-58s, later redesignated SH-34Gs) operated in pairs, with one helicopter searching for the target and the other armed for the kill. The S-61, which represented a major leap forward in design, first flew in prototype form on 11 March 1959. Deliveries of production HSS-2 Sea Kings to the Navy started in September 1961. In 1962 the designation was changed to SH-3A.

Sikorsky built 255 SH-3As, each powered by two 1,250-shp General Electric T58-GE-8B turboshaft engines. The follow-on U.S. Navy anti-submarine version was delivered from 1966 as the SH-3D Sea King, powered by 1,400-shp T58-GE-10 turboshaft engines, and it is still today its standard anti-submarine helicopter. Capable of 166 mph (267 km/h), it has provision for various weapons including torpedoes and depth charges. Conversion of SH-3As produced a small number of HH-3A

Below: US Navy Sikorsky SH-3H.

Above: VH-3D operated by the US Executive Flight Detachment as a VIP transport.

Navy search and rescue helicopters, while 105 SH-3As became SH-3Gs for utility duties, with a few for search and rescue in combat zones and carrying Minigun pods for defence. Some SH-3Gs were subsequently further modified into multi-purpose helicopters with updated anti-submarine equipment under the designation SH-3H.

From 1963 the Canadian Armed Forces received CH-124s, which are basically similar to SH-3As and were mostly assembled in Canada. Transport versions of the helicopter went to the Royal Danish Air Force and the Royal Malaysian Air Force as S-61As and S-61A-4 Nuris respectively. The SH-3D formed the basis for a very small number of S-61D-4s exported to Argentina and the U.S. Executive Flight Detachment received VH-3 VIP transports.

Apart from those assembled in Canada, Agusta of Italy also put the SH-3D into production under licence as an anti-submarine helicopter, following this model with the SH-3H and other versions for transport and rescue roles. Mitsubishi of Japan also produces versions for anti-submarine and rescue duties with the J.M.S.D.F. but the most important foreign licence-holder is Westland in the U.K. Continuing a tradition that goes back to the S-51, Westland concluded an agreement with Sikorsky to manufacture helicopters based on the SH-3D but with more powerful Rolls-Royce Gnome H.1400–1 turboshaft engines and equipment changes. Anti-submarine and search and rescue Westland Sea Kings have gone into service with the Royal Navy and the R.A.F. and have been exported. Westland also developed a tactical helicopter, based on the Sea King, which it called the Commando.

The Sikorsky S-61 also became an important helicopter airliner, the non-amphibious 30-passenger

Right: The Royal Australian Navy received ten Sea King Mk 50s from Westland for anti-submarine and other roles with No 817 Squadron. This Mk 50 uses its Plessey dipping sonar in an attempt to detect submarines.

04
NAVY

Top left: Westland Commandos ordered by the Saudi Arabian government for the Egyptian Air Force.

Left: An aviation first: a USAF CH-3C, the first version of the Sikorsky S-61R for military service, becomes the first helicopter ever to take on fuel from an aeroplane in flight, in January 1966.

S-61L and the amphibious 26/28-passenger S-61N being joined by a lightened version known as the Payloader for flying-crane operations. The Payloader is capable of lifting a 11,000-lb (4,990-kg) load in logging, construction and other heavy lifting operations.

In 1963 a further version of the S-61 appeared for the U.S.A.F. as the S-61R, with important new features that included an hydraulically operated rear ramp for straight-through loading of vehicles. Alternatively 30 troops, stretchers or cargo can be accommodated. The U.S.A.F. received two versions for transport and assault duties under CH-3 designations. Later, 50 CH-3Es were converted into HH-3Es for the U.S.A.F's Aerospace Rescue and Recovery Service. Taking the name Jolly Green Giants, these machines had new equipment which included light armament and armour protection, a retractable refuelling probe and a rescue hoist. Two Jolly Green Giants made the first-ever non-stop (flight refuelled) crossing of the Atlantic by helicopters between 31 May and 1 June 1967 while flying to the Paris Air Show. Similar rescue helicopters went to the U.S. Coast Guard as HH-3E Pelicans, carrying more advanced search equipment but without armament and armour.

Sikorsky S-61N amphibious airliner delivered to Bristow Helicopters.

Agusta HH-3F (Sikorsky S-61R) amphibian operated by the Italian Air Force.

In the year before the prototype S-61 flew, Sikorsky rolled out the amphibious S-62, a single-engined helicopter with the transmission and rotor system of the S-55 but with a fuselage of S-61 configuration though smaller. This became both a commercial airliner and a military search and rescue helicopter.

By scaling up the general configuration of the S-61, Sikorsky produced the S-65A which first appeared in 1964. Originally built with two 2,850-shp General Electric T64-GE-6 turboshaft engines and using components from the S-64 Skycrane, it entered U.S. Marine Corps service as the CH-53A Sea Stallion heavy assault transport. The Iranian Navy also received six. This version was followed by the more powerful CH-53D for the U.S.M.C., capable of carrying 55 troops, jeeps, missiles, a howitzer and carriage, or freight. Similar HH-53s went as armed helicopters to the U.S.A.F.'s Aerospace Rescue and Recovery Service. Exports were made to the armed forces of Austria, West Germany and Israel, for a variety of duties. Two HH-53Cs made a non-stop flight-refuelled crossing of the Pacific in August 1970. Also based on the CH-53 was the RH-53D (S-65MCM) mine countermeasures helicopter for the U.S. Navy. RH-53s, flying from the aircraft carrier USS *Nimitz*, took part in *Operation Evening Light*, the tragic and unsuccessful attempt to rescue American hostages held in Iran, on 24 April 1980.

A major increase in operational capability came with the delivery to the U.S.M.C. of Sikorsky CH-53E Super Stallions, helicopters which are currently the largest built anywhere other than in the Soviet Union. Powered by three T64-GE-416 engines of 3,696 shp maximum continuous rating each (making this also the most powerful non-Soviet helicopter), it is capable of 196 mph (315 km/h). A mine countermeasures version will join the U.S. Navy from 1986 as the MH-53E.

Sikorsky's other production activities centre of versions of the S-70 and S-76. The former was developed as a combat assault helicopter to supersede the Iroquois with the U.S. Army and carries 11 to 14 troops as the UH-60A Black Hawk. A naval version, known as the SH-60B Seahawk, is currently going into service on board U.S. Navy frigates as an anti-submarine and anti-ship surveillance and targeting helicopter. Seahawks are intended to complement the Navy's existing Kaman Seasprite LAMPS helicopters and not to replace them.

The S-76 is fundamentally a commercial and corporate helicopter with seating for 12 passen-

During the course of the US space programme, various helicopters from the S-58 have worked in the support role. Here a US Coast Guard HH-52A (Sikorsky S-62 variant) trains in the Gulf of Mexico in the recovery of Apollo 14 astronauts using the 'Billy Pugh' net.

MARINES

gers. Typical of the latest helicopters in this class – such as Bell's Model 222, the Italian Agusta A 109A and French Aérospatiale SA 365N Dauphin 2 – it has a retractable undercarriage to maintain the extremely clean fuselage lines in flight. Armed versions of the S-76 have also appeared. Sikorsky developed the S-76 in an attempt to take a larger share of the commercial helicopter market, a market dominated by Bell.

Perhaps the best known of all executive-type light helicopters is the Bell JetRanger, originally flown in 1962 as the Model 206 five-seater and currently available as the Model 206B JetRanger III. It was a JetRanger III that established another important milestone in the history of helicopters when, between 5 August 1982 and 22 July 1983, Australian Dick Smith flew *Australian Explorer* around the world, the staged journey covering 35,258 miles (56,742 km). This was the first-ever solo helicopter flight around the globe. Only on 30 September 1982 had the first-ever helicopter flight around the world ended, when Americans Ross Perot

Above: Sikorsky UH-60A Black Hawk helicopters despatch combat assault troops.

Opposite: The largest helicopter built outside the Soviet Union is the Sikorsky Super Stallion, operated by the US Navy/Marine Corps as the CH-53E and seen air-lifting the Grumman EA-6.

Above: ESSS (external stores support system) has been developed for the Black Hawk to allow the carriage of weapons and auxiliary fuel tanks. Each ESSS-equipped Black Hawk can thus carry over 10,000 lb (4,536 kg) of external stores, including sixteen Hellfire missiles as seen here during the flight qualification trials.

Left: Sikorsky SH-60B Seahawk, developed from the Black Hawk to fulfil the US Navy's requirements for a LAMPS III helicopter, flies over the guided missile frigate USS McInerney.

Right and overleaf: Sikorsky S-76 Mark II commercial helicopters used in support of offshore platforms. The Mark II followed the original model in 1982 and offers many refinements. Power for each helicopter is provided by two 650 shp Allison 250-C30S turboshafts.

Right: Agusta A 109A Mk II 6/7-passenger twin-turbine helicopter. This model is just one of several commercial and military versions produced.

Left: Bell Model 222B 7/9-passenger twin-turbine commercial helicopter, first flown in 1976.

Jr. and Jay. W. Coburn landed their LongRanger II *The Spirit of Texas* at Dallas after a 29-stage flight that had begun on 1 September. The Model 206L LongRanger is a stretched version of the JetRanger with accommodation for two extra passengers. Helicopters, similar in size to the JetRanger, are produced by several companies and include the Hughes Models 500 and 530, Hiller FH-1100, Aérospatiale Ecureuil and MBB BO 105.

Among other Bell products is a futuristic transport that can be flown in helicopter or aeroplane modes. Under development to a U.S. Army/NASA contract and designated XV-15 by the Army, it has a conventional aeroplane-type fuselage married to slightly forward-swept wings. At the tips of the wings are carried Avco Lycoming turboshaft engines driving 25-foot (7.62-metre) diameter rotors. With the engines tilted

Above: Aérospatiale SA 365N Dauphin 2, a 9/13-passenger twin-turbine commercial helicopter, just one of Aérospatiale's helicopters to use a 'fenestron' type ducted multi-blade fan as the anti-torque rotor. Another feature of the Dauphin 2 that is to be found in other modern Aérospatiale helicopters is the adoption of a Starflex rotor hub, in which the three conventional hinges (flapping etc) are superseded by maintenance-free balljoints of rubber and steel construction.

Left: An up-to-date Bell Model 206B JetRanger III flown by Rocky Mountain Helicopters for the news media.

Right: JetRanger III Australian Explorer *at the start of its round the world flight.*

Below: LongRanger III, developed in 1981 with a more powerful engine and increased fuel capacity than the LongRanger II. Power is provided by a 650 shp Allison 250-C30P, giving it a cruising speed of 126 mph (203 km/h).

Collins
Bell
NEWS

Above: LongRanger II The Spirit of Texas *making a difficult refuelling stop on board SS* President McKinley *in the North Pacific, in 40-knot winds, during the first helicopter round the world flight. The leg being flown at the time of this photograph was between Kushiro, Japan, and Shemya Island, Western Aleutians.*

Left: The pilots of the first helicopter round the world flight, Ross Perot Jr and Jay W. Coburn, at a return ceremony held in Dallas.

upwards and the rotors in horizontal position, the XV-15 can operate as a helicopter. With the engines facing forwards and the rotors acting as propellers, fast horizontal cruising flight is achieved. The operating advantages of such an aircraft in military and commercial forms are obvious but the concept of using rotors in this way is not new. In the 1960s Bell had designed and built a large transport-type research aircraft with four tilting propeller ducts, known as the X-22A, and, before that, a convertiplane known as the XV-3. Boeing's VZ-2A has already been mentioned, while others came from Curtiss-Wright and Hiller to name but a few.

Having had great military success with its Chinook, Boeing Vertol decided, in the 1970s, to develop a commercial version. The result was the Model 234 Commercial Chinook that first appeared in 1980. Available in three models, it can carry up to 44 passengers, and enormous external loads can be lifted by the LR (long range) and Utility versions, each at 28,000 lb (12,700 kg). Boeing Vertol's smaller Model 107 is currently only produced by Kawasaki in Japan in

A Phoenix Police five-seat Hughes Model 500 at sunset.

Right: The convenience and flexibility to business executives of operating a small helicopter is well illustrated by this Hughes Model 500E Olympian, a new version of the successful Model 500 series with cabin restyling to provide more leg and headroom.

Below: The Ecureuil is powered by a 641 shp Turboméca Arriel turboshaft engine. A version for the North American market was developed as the Avco Lycoming LTS 101-engined Astar. A 5/6 seat AS 350D Astar belonging to ERA Helicopters is illustrated, at rest in the Arctic region of Alaska.

Above: *The five-seat Hiller FH-1100 powered by a single Allison 250-C20B turboshaft engine. Maximum speed is 140 mph (225 km/h) and optional equipment can include a collapsible dual stretcher kit.*

Right: *Aérospatiale AS 350B Ecureuil Medevac version accommodating two stretchers across the rear cabin.*

Model II forms, for military, airline and various other duties.

The first Boeing Vertol Commercial Chinooks were ordered by British Airways, which wanted three, primarily for its support flights to oil rigs in the North Sea. Such large helicopters are unusual for this type of operation. Far more typical are single main rotor types like the Aérospatiale AS 332 Super Puma, the slightly larger and more powerful version of the original SA 330 Puma civil and military helicopter, with increases made to both accommodation and performance. Bristow Helicopters, a British operator and the largest in Europe with worldwide commitments, is one operator that uses Pumas and Super Pumas in support of offshore platforms. It was announced in May 1984 that the Super Puma has become the first helicopter in the world to be certificated for all-weather operations, including flight into icing conditions.

The Puma was one of three helicopters covered by a production agreement between Aérospatiale and Westland; the other two were

Above and overleaf: Two Bell helicopters operated commercially which use the Pratt & Whitney Aircraft of Canada PT6T-3B Turbo Twin Pac power plant are the 14 passenger or cargo Model 212 Twin Two-Twelve and similar Model 412. The Model 412 uses an advanced four-blade rotor and a new main rotor head with elastomeric bearings and dampers. The 412 is low on noise and vibration and has higher performance.

Opposite: MBB BO 105 CBS, a 5/6-seat light helicopter marketed in the USA as the Twin Jet II and powered by two 420 shp Allison 250-C20B turboshafts.

PK-HMU

Right: The XV-15 type of aircraft may have a bright future in military and commercial forms. The wing-tip 1,550 shp Avco Lycoming LTC1 K-4K turboshaft engines and 25 ft (7.62 m)-diameter rotors are being lowered here for conversion from vertical to horizontal flight.

Left: The tilt-rotor Bell XV-15 lands on board USS Tripoli. *It will perhaps become a familiar configuration for helicopters of the future.*

Below left: Flown for the first time in 1977 and developed originally as a military helicopter for Iran, the Bell Model 214ST SuperTransport is an 18 passenger or cargo commercial transport powered by two 1,625 shp General Electric CT7-2A turboshaft engines. It has a cruising speed of 159 mph (256 km/h) and the helicopter illustrated shows the optional wheeled undercarriage that became an option to the skid type in 1983.

Below: Boeing Vertol Model 234LR long-range variant of the Commercial Chinook for British Airways Helicopters.

1521
TRI-SERVICE
X-22A

Above: Like the British Airways Model 234LRs, those flown by Norway's Helikopter Service A/S can be converted rapidly from all-passenger to passenger and cargo 'combi' layouts.

Left: Bell's tilting-duct X-22A of 1966, powered by four 1,250 shp General Electric YT58-GE-8D turboshaft engines.

Right: Boeing Vertol 107 Model II manufactured in Japan by Kawasaki as a KV107II-4 tactical troop and cargo transport helicopter for the JGSDF, carrying extended-range auxiliary fuel tanks.

G-BFEU
BRISTOW

Left: SA 330J, the final civil version of the Aérospatiale Puma with two 1,575 shp Turboméca Turmo IVC turboshaft engines, accommodating up to 20 passengers.

Below: The first production helicopter with a 'fenestron' thirteen-blade ducted anti-torque fan was the Aérospatiale SA 341/342 Gazelle, an RAF Gazelle HT.Mk 3 training version (SA 341D) shown here.

the Gazelle and the Lynx. All three have another factor in common: each was primarily intended for service with the forces of France and the U.K. but have also been exported. However the Lynx, which is the only one of the three in which Westland has design leadership, is also the only helicopter of the trio not built in civil form. Instead, Westland has developed its own twin-engined and enlarged derivative as the Westland 30. It is the company's first totally new helicopter (excluding the Lynx) since the Wasp and Scout. Intended for civil and military use, the first production Westland 30s went to commercial operators in 1982.

Of the helicopters that took part in the Falklands conflict of 1982, the Westland Lynx and Sea King gained most headlines. On 25 April, the very first day of British

Left: The Royal Netherlands Navy has received Lynx Mk 25s (UH-14A in service) for search and rescue and Mk 27s (SH-14Bs) for anti-submarine warfare.

Overleaf: Westland 30 Series 100 17-passenger commercial helicopter delivered to Airspur, the first foreign customer, in 1982. Two 1,135 shp Rolls-Royce Gem 41-1 turboshaft engines.

airspur

Right: Mock-up of the latest version of the Westland Lynx, the Lynx 3 anti-tank helicopter with a mast-mounted sight and provision for missiles including Hot, TOW and Hellfire.

Below: Lynx HAS.Mk 2 during deck handling trials on board HMS Birmingham. *It was two such helicopters armed with Sea Skua missiles that damaged the Argentine patrol ship* Alferez Sobral *and sank* Comodoro Somellera *during the Falklands conflict of 1982.*

Task Force action, Lynx helicopters from HMS *Alacrity* and *Antelope* crippled the Argentine submarine *Santa Fe* off Grytviken harbour, South Georgia. Later in the same day Royal Marines were carried onto South Georgia by Sea Kings. Other actions followed and it is sobering to realize that, on the day of 21 May alone, a detachment of Sea Kings from No. 846 Squadron managed to air-lift over 407 tons of supplies and 520 troops during the British landings at Port San Carlos in the East Falklands. It was the lessons learned from this war that led Westland to develop an early-warning version of the Sea King, carrying Searchwater marine surveillance radar in an inflatable container which is swung down during deployment. The first Sea King AEWs put to sea on HMS *Illustrious* in August 1982.

Right: After the Falklands experience, it took just eleven weeks to develop an early warning version of the Sea King equipped with Searchwater marine surveillance radar.

ROYAL NAVY
ROYAL NAVY
88
CU

Largest and fastest

Chapter Five mentions the Kamov Ka-25 (NATO *Hormone*), which became the standard anti-submarine helicopter of the Soviet Navy on board many classes of warship and from shore bases, superseding the Mil Mi-4. Production lasted from 1966 to 1975. Typically a Kamov type as witnessed by its adoption of co-axial contra-rotating rotors, driven by two 900-shp Glushenkov GTD-3 turboshaft engines, it was also exported to India, Syria and Yugoslavia (some, if not all, were ex-Soviet Navy helicopters). A civil flying-crane derivative, the Ka-25K, can lift a 4,410-lb (2,000-kg) payload.

In the early 1980s a new Kamov helicopter of larger proportions appeared but still with co-axial twin rotors. This is said to be the Ka-27 in military anti-submarine form and Ka-32 in civil form. The Ka-27 is also capable of performing

A new Kamov Helix-A *on the helicopter pad of the Soviet guided missile destroyer* Udalov *sailing in the Baltic.*

Kamov Ka-26 used for agricultural work in Japan by the Asahi Helicopter Company. Note the open area to the rear of the crew cabin.

to new standards another of the Ka-25's tasks, that of locating tactical over-the-horizon targets for Soviet ships and aircraft which are carrying cruise missiles and of correcting the launched missile's mid-course flight should it be off target. Known to NATO as *Helix*, the Ka-27 has joined a new class of Soviet guided-missile destroyers.

A much smaller Kamov helicopter of the mid-1960s is the Ka-26 (NATO *Hoodlum*), which is the smallest production helicopter to adopt a layout in which the space to the rear of the flight cabin can be occupied by detachable payloads. This arrangement, however, has made the Ka-26 ideally suited to many specialized tasks, including flying-crane, agricultural spraying and dusting, and firefighting. Furthermore, with a cabin pod attached, the Ka-26 can carry six passengers, stretchers or internal cargo.

At the other end of the scale, Mil has produced both the world's fastest helicopter in the form of the Mi-24 (previously described) and the world's largest. Having abandoned its 1960s-built V-12, a giant twin-rotor helicopter with an overall rotor-tip to rotor-tip span of 219 feet 10 inches (67 metres) and powered by four 6,500-shp Soloviev D-25VF turboshaft engines, Mil set about the task of designing a more practical heavy transport helicopter. The resulting Mi-26 (NATO *Halo*) first flew in about 1979. With the exception of the

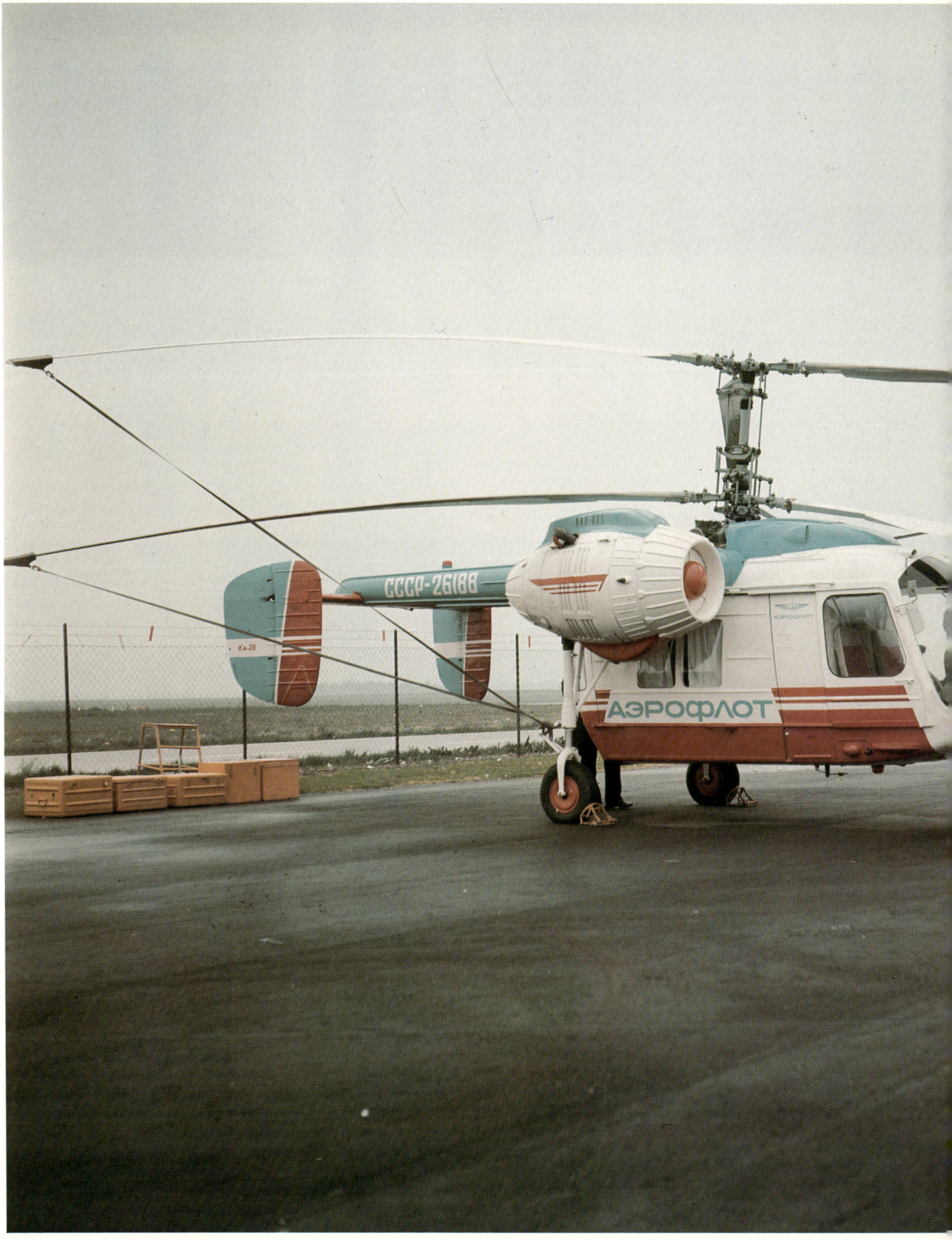
CCCP-26188
АЭРОФЛОТ

abandoned V-12, the Mi-26 is the heaviest helicopter ever flown, weighing about 123,450 lb (56,000 kg). Powered by two 11,400-shp Lotarev D-136 free-turbine turboshaft engines, driving a single eight-blade rotor, it has a main freight cabin of 39 feet 4 inches (12 metres) in length with the rear loading door closed. Its maximum payload is 44,090 lb (20,000 kg), which gives it a capacity equal to that of a Lockheed C-130 Hercules transport aeroplane. Instead of cargo, which can include large vehicles, the Mi-26, in military form, can accommodate 100 or more troops.

Previous chapters have mentioned world records currently held by rotary-winged aircraft. These have encompassed Wallis autogyros, the Soviet Ka-22 and British Fairey Rotodyne convertiplanes, the French Aérospatiale Lama and the Soviet A-10 helicopters, the latter two holding the current height and speed records respectively. One other record that begs inclusion is distance in a straight line flown by a helicopter. This record is held by a Hughes OH-6A at 2,213 miles (3,561.55 km), set during 6–7 April 1966.

Above: The largest helicopter ever built was the Mil M-12, which on 6 August 1969 carried an 88,635.64 lb (40,204.5 kg) payload to an altitude of 2,000 metres.

Left: Aeroflot-operated Ka-26 with a 6-passenger cabin pod attached.

Why Leave it to the Professionals?

Simple autogyros have given more people the pleasure of owning and flying rotary-winged aircraft than helicopters. The fact that they are cheaper to purchase in ready-to-fly form, somewhat easier to construct from bought plans or kits, and simpler to design for those with the technical knowledge to attempt such a feat, could have led to no other situation. Yet those who have the resources and skills to go beyond the autogyro to the more versatile helicopter can enjoy flying experiences beyond those available to autogyro pilots. To the few that can actually design, build and fly their own helicopters, this pleasure is multiplied many times.

All pioneer helicopter designers were amateurs in this field at one stage or another and it was only through their experiments and determination that the practical helicopter emerged. By the 1950s, however, with helicopters and helicopter manufacturers becoming established, the first simple helicopters, intended for the amateur constructor market, appeared. Among the first of these was the American Adams-Wilson XH-1 HobbyCopter, an open structure single-seater powered by a 40-hp Triumph motorcycle engine. This made its maiden flight in November 1958 and within a year more than 100 sets of plans had been sold, together with kits of components to enthusiasts worldwide.

The United States remains the main centre for 'homebuilt' aircraft of all types, although in several other countries helicopters designed and built by their owners have appeared. Probably the best known U.S. company currently offering kits of component parts to would-be owner/pilots is RotorWay Aircraft Incorporated of Arizona, which offers the Scorpion 133 and Exec two-seaters, both powered by the 145-hp RotorWay RW-145 piston engine. The Exec is capable of 112 mph (180 km/h). The Scorpion 133 has a particularly interesting history, having previously been known as the Scorpion Too, which was itself developed from the single-seat Scorpion, the original helicopter offered in plans and kit form by RotorWay Aircraft. However, the Scorpion was, in fact, the work of B. J. Schramm, who had established the Schramm Company in 1958 to market, in kit form and as a finished helicopter, a single-seater known as the Javelin.

Another source of helicopter kits in the U.S.A. is Tamarind International of Louisiana, which offers the single-seat Commuter IB, the two-seat Commuter IIA and the roomier Commuter IIB. The two-seat Commuters, with 150-hp Avco Lycoming O-320 engines, were previously marketed by International Helicopters Incorporated, which had based its helicopters on those once offered for sale by Helicom Incorporated. Helicom first marketed single- and two-seat helicopters during the 1960s.

Right: Currently the largest helicopter in the world is the Soviet Mil Mi-26.

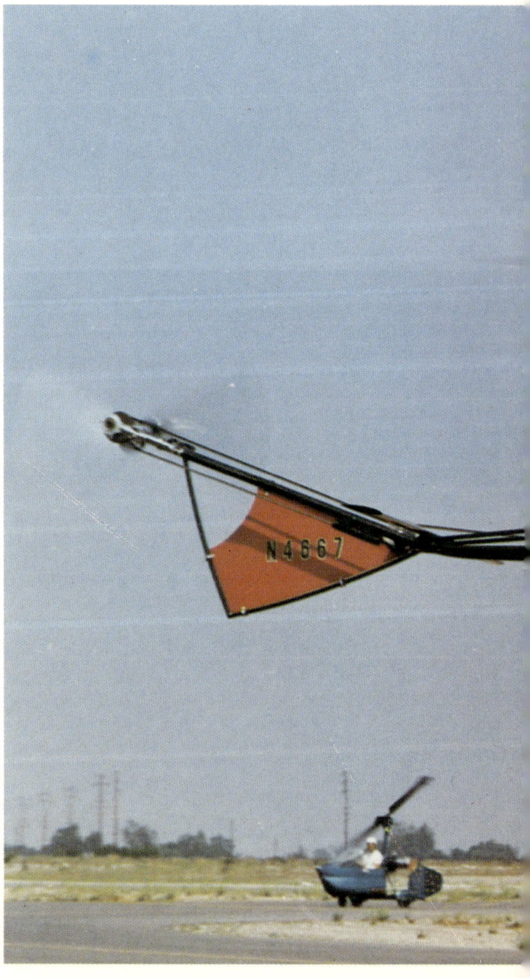

Right: RotorWay Scorpion 133, a popular helicopter for home assembly.

СССР-06141
Н-351
АЭРОФЛОТ

The Way Forward

The helicopter, so very difficult to develop into a successful and useful form of aircraft and now so much a part of everyday life, has come a long way since Frenchman Paul Cornu managed to hover one foot above the ground for 20 seconds in 1907. The first production helicopters from Sikorsky went to the U.S. Army Air Force. In 1948 the U.S. Army Air Force became the U.S. Air Force, a separate force to the U.S. Army, leaving the Army only transport and liaison aircraft. The helicopter, so suited to army-style operations, gradually introduced new firepower and today the U.S. Army is almost certainly the operator of the largest number of helicopters, totalling around 9,000 of all kinds including gunships. The air force with the greatest helicopter firepower belongs to the Soviet Union, which also operates the largest helicopters for transport duties.

What does the future hold in store for helicopters? The day of the helicopter air-to-air fighter is approaching but, in a wider context, aircraft of many sizes that combine the best attributes of aeroplanes and helicopters, as demonstrated by the Bell XV-15, may gain a strong foothold in VTOL operations. This type of aircraft combines the helicopter's vertical flying abilities with the aeroplane's cruising speeds but, like the helicopter, will remain more expensive to purchase and operate than a conventional aeroplane.

Index

Figures in italics indicate the inclusion of illustrations